International Science and Technology Conference for Youth "Advanced Materials for Engineering and Medicine" (AMEM-2019)

IOP Conference Series: Materials Science and Engineering
Volume 731

Tomsk, Russia
30 September - 5 October 2019

ISBN: 978-1-7138-2590-6
ISSN: 1757-8981

Printed from e-media with permission by:

Curran Associates, Inc.
57 Morehouse Lane
Red Hook, NY 12571

Some format issues inherent in the e-media version may also appear in this print version.

Institute of Physics
Dirac House, Temple Back
Bristol BS1 6BE UK

Phone: 44 1 17 929 7481
Fax: 44 1 17 920 0979

techtracking@iop.org

Additional copies of this publication are available from:

Curran Associates, Inc.
57 Morehouse Lane
Red Hook, NY 12571 USA
Phone: 845-758-0400
Fax: 845-758-2633
Email: curran@proceedings.com
Web: www.proceedings.com

TABLE OF CONTENTS

INVESTIGATION OF THE STRUCTURE AND PROPERTIES OF SWCNT-MODIFIED EPOXY SPECIMENS 1
M V Burkov, A V Eremin

STRUCTURE AND PROPERTIES OF SWCNT-MODIFIED CFRP 6
M V Burkov, A V Eremin

ALGORITHM FOR J-INTEGRAL MEASUREMENTS BY DIGITAL IMAGE CORRELATION 11
V V Titkov, A V Eremin, S V Panin, P S Lyubutin, A A Kozulin

DEFORMATION BEHAVIOR UNDER STATIC AND CYCLIC TENSION OF POLYMER GRAFTS WITHOUT AND AFTER MODIFICATION BY RGD PEPTIDES 16
S V Panin, L V Antonova, A V Byakov, T V Glushkova, E O Krivkina, A V Eremin, V N Silnikov, L S Barbarash

APPLICATION OF LASER WELDING IN CAR BODIES MANUFACTURING 24
J Viňáš, J Brezinová, J Brezina, P Maruschak, S Panin

THE EFFECT OF PLASTIC DEFORMATION ON THE GEOMETRIC PARAMETERS OF NB-TI-BASED SUPERCONDUCTING CABLE 30
A Yntymakova, G Shlyakhova

COMPARATIVE ANALYSIS OF THE POLYMERIC MATERIALS DEFORMATION BEHAVIOR UNDER SQUEEZED AND FREE COMPRESSION 35
A A Adamov, A A Kamenskikh, Yu O Nosov

RESEARCH OF THE MATERIALS AND REINFORCEMENT SCHEMES EFFECTS ON A MECHANICAL BEHAVIOR SMART STRUCTURE WITH EMBEDDED PIEZO ACTUATORS 41
A N Anoshkin, P V Pisarev, V A Ashikhmin

THE INFLUENCE OF THE THREE-LAYER MOUTHGUARD ON THE STRESS-STRAIN STATE OF A PAIR OF OPPOSING TEETH 47
E S Avdeeva, A A Kamenskikh

FRICTION INDUCED STRUCTURAL CHANGES OF MANGALLOY OBSERVED BY SYNCHROTRON X-RAY DIFFRACTION AND METALLOGRAPHIC ANALYSIS 53
K I Emurlaev, A. A. Ruktuev, A E Karmanova, V G Burov, A A Nikulina, I A Bataev

INFLUENCE OF A DELAMINATION DEFECT LOCATION ON THE OPERATING LIFE OF A MULTI-LAYERED COMPOSITE SAMPLE 59
A N Anoshkin, P V Pisarev, D A Ermakov

INFLUENCE OF A DELAMINATION TYPE DEFECT ON THE OPERATIONAL LIFE OF A SOUND-ABSORBING STRUCTURE WITH A HONEYCOMB FILLER 65
A N Anoshkin, P V Pisarev, D A Ermakov

NANOSTRUCTURING OF A SURFACE LAYER AS A WAY TO IMPROVE THE MECHANICAL PROPERTIES OF HYPOEUTECTIC SILUMIN 70
A A Kondratyuk, Yu F Ivanov, A A Klopotov, A M Ustinov, Yu A Abzaev, E A Petrikova, A D Teresov, A S Tolkachev

ELECTROMAGNETIC PROPERTIES OF COMPOSITE MATERIALS BASED ON ABS PLASTIC WITH CARBON NANOTUBES OBTAINED BY THE ADDITIVE TECHNOLOGY IN THE SHF AND EHF BANDS 76
G E Kuleshov, A V Badin, M O Gering

IMPROVING THE MECHANICAL PROPERTIES OF SIC-CERAMICS BY MEANS OF VACUUM ELECTRON-ION-PLASMA ALLOYING WITH TITANIUM 82
A A Leonov, Yu F Ivanov, M P Kalashnikov, E V Abdulmenova, V V Shugurov, A D Teresov

ELECTROMAGNETIC PROPERTIES OF HEXAFERRITE POLYMER COMPOSITE MATERIALS 89
A V Sbrodov

INFLUENCE OF HETEROGENEITY COMPOSITE FILAMENT FOR ADDITIVE TECHNOLOGY ON ITS ELECTROPHYSICAL PROPERTIES AT THZ RANGE 94
P P Smygalina, A V Badin, D A Pidotova, E A Trofimov

QUASI-OPTICAL RESONANT SINGLE-FREQUENCY FLAW DETECTOR OF POLYMER FILAMENT FOR 3D PRINTING 100
D D Teterina, I O Dorofeev, A V Badin

INCREASING FATIGUE LIFE OF 09MN2SI STEEL BY MEANS OF HIGH-TEMPERATURE MULTISTEP HELICAL ROLLING 105
I V Vlasov, N S Surikova, S V Panin, A V Yakovlev, J F Gomorova

CYTOTOXIC EFFECTS OF GRANULATED HYDROXYAPATITE WITH VARIOUS PARTICLE SIZE DISTRIBUTION 112
O Zakharova, A Gusev, K Chuprunov, A Yudin, D Kuznetsov

NONWOVEN PTFE MEMBRANES FABRICATED BY ELECTROSPINNING METHOD: PREPARATION AND CHARACTERIZATION 118
I M Kolesnik, E N Bolbasov, V M Buznik, S I Tverdokhlebov

COMPOSITION POLYMERIC MEMBRANES BASED ON THE VDF-TEFE COPOLYMER FORMED BY ELECTROSPINNING 124
T S Tverdokhlebova, E N Bolbasov, V M Bouznik

THE PH LEVEL INFLUENCE ON HYDROXYAPATITE PHASE COMPOSITION SYNTHESIZED WITH HYDROTHERMAL METHOD 130
K Chuprunov, A Yudin, D Lysov, E Kolesnikov, D Kuznetsov, D Leybo, I Ilinykh, A Godymchuk

THE HYDROTHERMAL SYNTHESIS DURATION INFLUENCE ON CALCIUM PHOSPHATE AND HYDROXYAPATITE PHASE COMPOSITION 137
K Chuprunov, A Yudin, D Leybo, D Lysov, E Kolesnikov, D Kuznetsov, A Godymchuk, A R Mandal

EFFECT OF PARAMETERS OF ULTRASONIC SHOT PEENING ON THE PHYSICOMECHANICAL PROPERTIES OF THE SURFACE LAYER OF FLAT SAMPLES 144
A A Shiryaev, V N Trofimov, V V Karmanov, D V Vindokurov

RESEARCH METHODOLOGY OF AMPLITUDE-FREQUENCY CHARACTERISTICS OF DENTAL (ENDODONTIC) FILES 150
A V Avilov, E V Laktionov

Author Index

IOP Conf. Series: Materials Science and Engineering **731** (2020) 011001 doi:10.1088/1757-899X/731/1/011001

International science and technology conference for youth "Advanced Materials for Engineering and Medicine" (AMEM-2019)

30 September – 5 October 2019, Tomsk, Russia

The International science and technology conference for youth "Advanced Materials for Engineering and Medicine" (AMEM-2019) was held at Tomsk Polytechnic University (TPU) in 30 September – 5 October 2019.

The scientific program consisted of plenary sessions and parallel sessions and included the following topics:

- Problems of materials reliability
- Functional materials
- Surface and coating modification
- Advanced materials for medicine

We want to thank all members of the Organizing Committee, TPU and everyone who contributed to the organization of this event, for their contribution towards the conference and for their contributions to this proceeding.

The Organizing Committee of AMEM-2019

IOP Conf. Series: Materials Science and Engineering **731** (2020) 011002 doi:10.1088/1757-899X/731/1/011002

Peer review statement

All papers published in this volume of *IOP Conference Series: Materials Science and Engineering* have been peer reviewed through processes administered by the proceedings Editors. Reviews were conducted by expert referees to the professional and scientific standards expected of a proceedings journal published by IOP Publishing.

IOP Conf. Series: Materials Science and Engineering **731** (2020) 012001 doi:10.1088/1757-899X/731/1/012001

Investigation of the Structure and Properties of SWCNT-Modified Epoxy Specimens

Burkov M V[1,2] and Eremin A V[1]

[1]Institute of strength physics and materials science, Akademicheskiy avenue 2/4, Tomsk, 634055, Russia
[2]Tomsk Polytechnic University, Lenin avenue 30, Tomsk, 634050, Russia

burkovispms@mail.ru

Abstract. SWCNT-modified epoxy with advanced electrical properties was investigated in the paper. CFRPs have high specific strength and elastic modulus which are the most important properties for aerospace engineering. However a lack of electrical and heat conductivity makes efficient lightning strike protection and thermal de-icing complicated. In this research electric conductivity of epoxy binder for CFRP production is modified by addition of single-wall carbon nanotubes. Measurements of electrical conductivity of hardened epoxy binders are performed showing good increase of conductive properties.

1. Introduction

Nowadays the application of composites has increased drastically. Being the materials composed of two or more constituents they provide an unlimited variety of physical and mechanical properties that fit the exact application demands. One of the modern trends in the field of engineering of composites for structural applications is related to combined reinforcing by fiber-based materials (fabrics, mats, etc) and different additives (microfibers, nanofillers, etc). This allows obtaining better mechanical properties and achieving new quality that is not inherent to traditional fiber-reinforced polymers (FRP). For example one approach for providing electrical conductivity to carbon FRP is addition of carbon nanotubes (CNT) resulting in so-called hybrid CFRP [1,2].

There are different additives being studied to modify FRPs. Metal powders can be used for providing electrical conductivity and improving thermal conductive properties [1,3,4]. Ferromagnetic particles are applied for the production of electromagnetic shielding composites [1,5]. Different research groups use microsized additives and nanomaterials [6,7] concluding that both types can be used for manufacturing of hybrid FRPs having advantages and disadvantages.

One of the technological problems which should be solved when designing a hybrid CFRP is the homogeneous distribution of small-sized reinforcement in the matrix (thermoset or thermoplastic). In the case of microfillers like short milled fibers of metal powders this problem is often solved by using such techniques like manual stirring or mechanical mixing. But as for nanomaterials the dispersing of particles is quite difficult resulting in the necessity of application of high-energy mixing procedures [8,9]. One of the robust techniques is ultrasound (US) assisted mixing, which can be used both in laboratory conditions and in industrial environment. There are a lot of scientists investigating US treatment for production of different advanced materials.

The technology of mixture preparation is often limited only to laboratory conditions and equipment resulting in high costs and low productivity. Thus it is quite important to search for novel methods that will be both efficient and cost-effective. The paper is devoted to the investigation of structural features and mechanical properties of SWCNT-modified epoxy specimens obtained under different conditions.

2. Materials and methods

The specimens were prepared in order to investigate structural properties, compressive strengthand electrical conductivity. Single-wall carbon nanotubes used are provided by OcSiAl (Russia, Novosibirsk) having an outer mean diameter of 1.6±0.4 nm and a length of more than 5 μm. The residual impurities and moisture are less than 15 and 5 wt.% correspondingly. Epoxy resin of L grade and hardener CL were supplied by R&G, Germany. This epoxy binder (L epoxy + CL hardener) can be used to produce medium-to-high performance CFRPs which have the properties comparable to CFRP made using aerospace-grade epoxies.

The mixtures of epoxy and SWCNT were prepared by mechanical mixing during 1 minute and ultrasonication using UZDN-A (Russia) probe sonicator for 20 minutes with a frequency of 22 kHz and power of 130 W. During ultrasonic mixing the glass was constantly cooled in water. After cooling the hardener was added to the mixture and poured into molds for preparation of cylindrical specimens with a diameter of 12.7 mm and a length of 30 mm. Summary there were initial non-modified epoxy specimens and 3 types of SWCNT-modified with 0.1 wt.%, 0.2 wt.%, and 0.3 wt.% of SWCNT. In order to perform conductivity measurement the specimens were coated with silver conductive paint. The paint was applied on the ends of the cylinder and electrical conductivity was measured between these pads using LCR meter E7-20, Russia.

After measurement of electrical properties the specimens were tested by uniaxial compression using Instron 5582 universal testing machine at the rate of 2.5 mm/min. Three specimens of each CNT content were prepared. The fracture surfaces of the specimens were investigated by SEM using LEO EVO 50 (Zeiss, Germany) in “Nanotech” ISPMS SB RAS.

3. Experimental results and discussion

3.1. Electrical conductivity

The results of conductivity measurement are presented in Figure 1 and Table 1. Non-modified epoxy has nearly zero electrical conductivity: the upper threshold of resistance measurement of the used device is 10^9 Ohm while the measured resistance exceeds this threshold resulting in electrical conductivity less than 10^{-7} S/m.

The addition of 0.1% of SWCNTs increases the conductivity to 0.012 S/m. Larger content of SWCNTs results in nonlinear (Figure 1) rise of conductivity to 0.063 S/m for specimen CNT-3. The scatter of the conductivity for CNT-1 is about 6% while for CNT-2 – 27%. The obtained results demonstrate the possibility of enhancing the electrical conductivity of different materials using SWCNTs: the dielectric epoxy after modification acquired conductive properties.

Table 1. Electrical conductivity of specimens.

Specimen	Electrical conductivity, S/m
CNT-0	$<10^{-7}$
CNT-1	0.012±0.0007
CNT-2	0.03±0.0083
CNT-3	0.063±0.0025

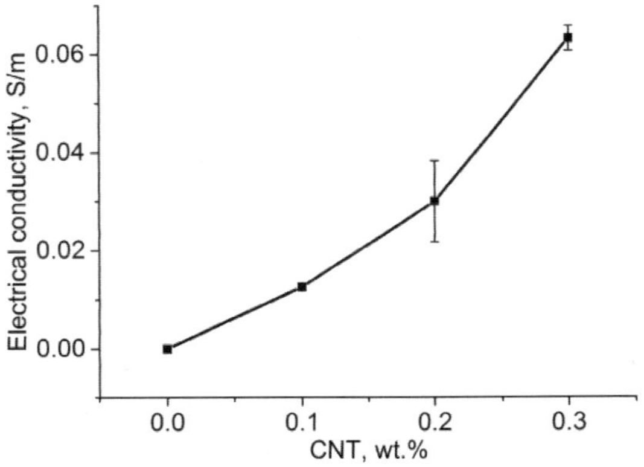

Figure 1. Results of conductivity measurements of epoxy specimens.

3.2. Compression testing

Because the viscosity of epoxy due to addition of SWCNTs drastically increases thus final porosity of epoxy specimens with SWCNTs has risen. The unmodified epoxy has low porosity and compression strength of ~111 MPa. The addition of 0.1% and 0.2% of SWCNTs results in decrease of compression strength to 87 and 63 MPa consequently. For specimen CNT-3 it was decided to use degassing procedure which resulted in decrease of porosity leading to better compressive strength – 105 MPa.

3.3. Scanning electron microscopy

Figures 2, 3 and 4 show SEM images of fractured epoxy specimens CNT-1, CNT-2, and CNT-3 respectively. It can be seen from these results that the specimen CNT-1 fractures with mainly viscous behavior. The porosity of the specimen is low, but compared to the non-modified epoxy there is a drop in compressive strength (87 MPa versus 111 MPa) which leads to the emergence of pores.

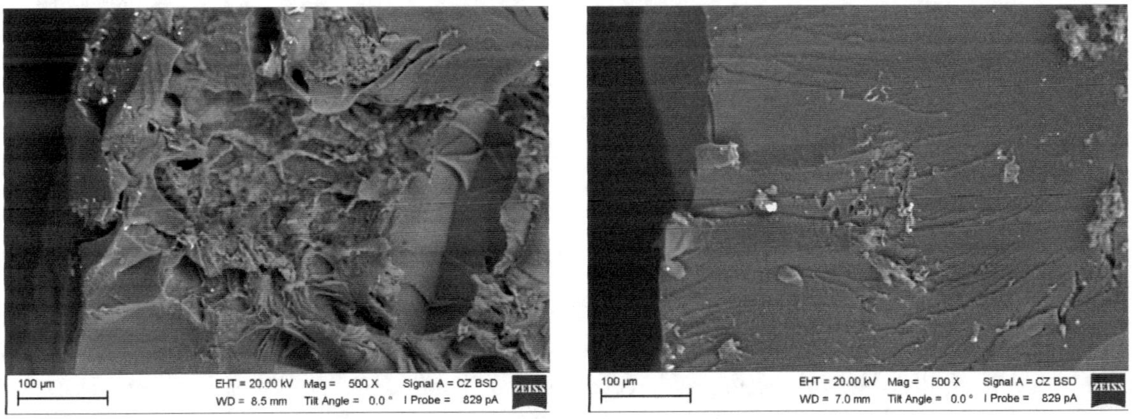

Figure 2. SEM images of fractured CNT-1 specimen.

Larger content of SWCNTs of the CNT-2 specimen results in the remarkable increase in viscosity making it difficult to homogeneously mix the epoxy with hardener without the introduction of air in the mixture. Small air bubbles combine in large pores during molding and there are easily seen in the fracture surfaces of the CNT-2 specimen (Figure 3). The fracture behavior transits to the more brittle one and along with porosity the mean compressive strength drops to 63 MPa.

Figure 3. SEM images of fractured CNT-2 specimen.

As can be seen from the SEM images of the CNT-3 specimen the structure contains fewer pores than the structure of CNT-1. It is due to the use of a degassing procedure which resulted in decrease of porosity and preservation of compressive strength at a level of neat epoxy (105 MPa). The fracture behavior is even more brittle.

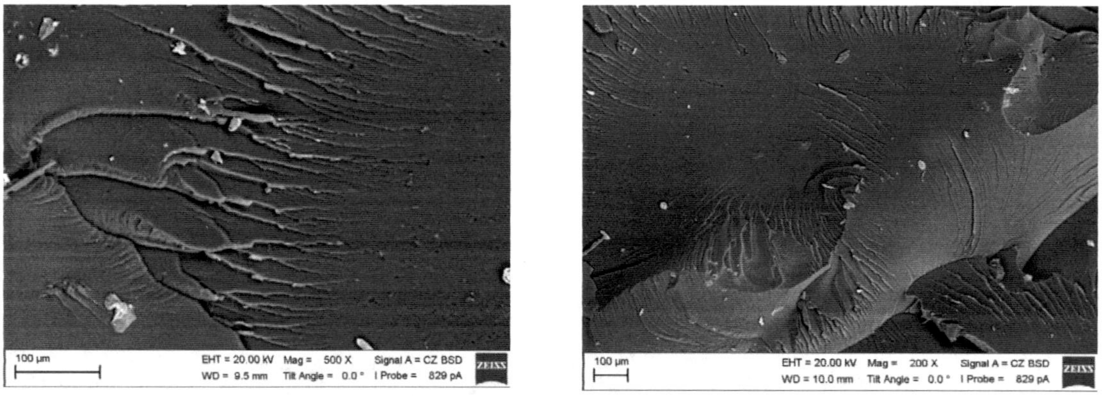

Figure 4. SEM images of fractured CNT-3 specimen.

4. Conclusion

The SWCNT modified epoxy was investigated. The research was focused on the evaluation of electrical conductivity and compressive properties. Electrical conductivity after the addition of SWCNT rose from null values to 0.063 S/m. It is found that the viscosity during addition of SWCNT increases drastically thus the degassing procedure is obligatory. Otherwise significant porosity is emerged. The method proposed in the work for producing hybrid composites has good potential and future research is to be linked to the development of reliable and inexpensive preparation and mixing techniques for carbon fiber reinforced polymers.

Acknowledgements

The research was supported by Grant of Russian Science Foundation №19-79-10148.

References

[1] Gibson R F 2010 A review of recent research on mechanics of multifunctional composite materials and structures *Compos. Struct.* **92** 2793–810

[2] Tehrani M, Safdari M, Boroujeni A Y, Razavi Z, Case S W, Dahmen K, Garmestani H and Al-Haik M S 2013 Hybrid carbon fiber/carbon nanotube composites for structural damping applications *Nanotechnology* **24** 155704

[3] Cortes L Q, Racagel S, Lonjon A, Dantras E and Lacabanne C 2016 Electrically conductive carbon fiber / PEKK / silver nanowires multifunctional composites *Compos. Sci. Technol.* **137** 159–66

[4] Jang J, Park H C, Lee H S, Khil M-S and Kim S Y 2018 Electrically and Thermally Conductive Carbon Fibre Fabric Reinforced Polymer Composites Based on Nanocarbons and an In-situ Polymerizable Cyclic Oligoester *Sci. Rep.* **8** 7659

[5] Zhao Z J, Zhang B Y, Du Y, Hei Y W, Yi X S, Shi F H and Xian G J 2017 MWCNT modified structure-conductive composite and its electromagnetic shielding behavior *Compos. Part B Eng.* **130** 21–7

[6] Panin S V., Duc Anh N, Kornienko L A, Alexenko V O, Buslovich D G and Ovechkin B B 2018 Wear-resistant polyetheretherketone composites *Mater. Today Proc.* **5** 25976–82

[7] Panin S V., Le T M H, Kornienko L A, Alexenko V O, Ivanova L R and Ovechkin B B 2018 Mechanical and tribotechnical properties of polyphenylene sulfide composites reinforced with carbon fibers of various dimension p 020231

[8] Caneba G T, Dutta C, Agrawal V and Rao M 2010 Novel Ultrasonic Dispersion of Carbon Nanotubes *J. Miner. Mater. Charact. Eng.* **09** 165–81

[9] Kim M and Kim J 2018 Nanoparticle dispersionizer by ultrasonic cavitation and streaming *Jpn. J. Appl. Phys.* **57** 07LE03

IOP Conf. Series: Materials Science and Engineering **731** (2020) 012002 doi:10.1088/1757-899X/731/1/012002

Structure and Properties of SWCNT-Modified CFRP

Burkov M V[1,2] and Eremin A V[1]

[1]Institute of strength physics and materials science, Akademicheskiy avenue 2/4, Tomsk, 634055, Russia
[2]Tomsk Polytechnic University, Lenin avenue 30, Tomsk, 634050, Russia

burkovispms@mail.ru

Abstract. Structure and mechanical properties of CFRP modified by SWCNT with different content (0.1, 0.2 and 0.3 wt. %) have been studied in the present paper. The structural investigations include SEM imaging while the mechanical properties have been characterized via tensile and flexural tests. The main functional feature of such hybrid composite is an electrical conductivity, which has been also measured. Out-of-plane conductivity has been enhanced up to ~66 S/m from dielectric unmodified CFRP. It has been shown that tensile properties stay almost unchanged, while flexural strength and modulus have increased by 22% and 16% correspondingly.

1. Introduction

In recent decades the application of composites in the industry was significantly increased due to their superior mechanical properties as well as an opportunity to obtain wide range of variation of these properties by producing composites with different consistuents and contents in order to satisfy all requirements. In this case, the new trend in the composite manufacturing is hybrid composites, which together with conventional continuous fiber reinforcement applies dispersive nano- or micro- fillers. Due to such a technique it is possible to produce materials with completely new properties or enhance existing.

Concerning carbon fiber-reinforced composites (CFRP) used in the aerospace industry, a very promising way is an addition of single-wall (SW) or multi-wall (MW) carbon nanotubes (CNT) [1,2]. CNTs added to CFRP results in an increase in composite of electrical and heat conductivity, which could be useful for lightning protection and higher efficiency of de-icing systems. Similar results with some electromagnetic shielding effect could be obtained using metal powders [1,3,4,5]. Nanomaterials are much more expensive than microfillers, but it is shown in papers that microsized additives [6,7] can be successfully used for the manufacturing of hybrid fiber-reinforced polymers.

The described approach has a limitation in its industrial application due to a problem in a homogeneous dispersing of small-sized reinforcement in the matrix. One of the ways for solving the issue is ultrasound (US) assisted mixing, applicable in the laboratory or industrial environment [8,9].

The present paper is devoted to the laboratory investigation of the structure and the properties of CFRP filled with SWCNT using ultrasonic homogenization and manual layup.

2. Materials and methods

Hybrid CFRP/SWCNT specimens were prepared in order to investigate tensile and flexural strength and electrical conductivity. Single-wall carbon nanotubes used are provided by OcSiAl (Russia,

Novosibirsk). SWCNT had a mean diameter of 1.6±0.4 nm and a length of more than 5 μm. Epoxy resin of L grade and hardener CL were manufactured by R&G, Germany. This epoxy binder (L epoxy + CL hardener) can be used to produce medium-to-high performance CFRPs which have the properties comparable to CFRP made using aerospace approved binders. The carbon reinforcement used is 200 g/m^2 dense plain fabrics of Tenax® HTA 40/200 tex, 3k carbon fiber. VATI basalt plain fabrics of 200 g/m^2 were used in the middle of the stack allow avoiding the contact between neighboring carbon fiber layers and helping to investigate directly the effect of CNTs on electrical conductivity. The stacking sequence of $[(0/90_C)_4,(0/90_B)_2]_S$ results in a 2.1 mm thick 12-ply orthotropic laminate. Hand lay-up with hot pressing using Gotech 7014-R (pressure 0.48 MPa, heating of plates to 60 °C, and 12 hours post-cure) was used to fabricate blanks which were cut into specimens using milling machine. Table 1 and Figure 1 present the specimens investigated in the research.

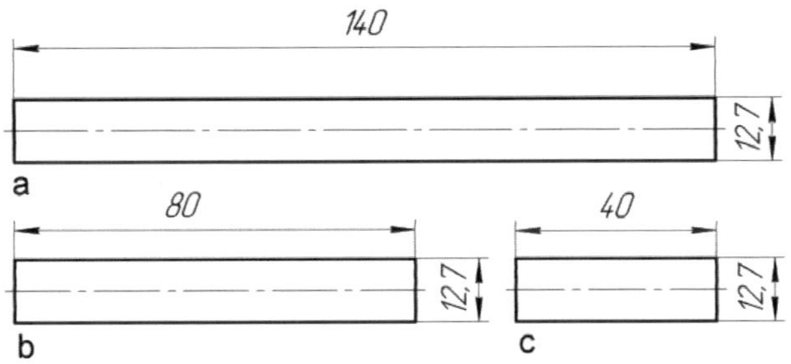

Figure 1. Sketches of the specimens for a) tension; b) 3-point bending; c) conductivity measurement.

The mixtures of epoxy and SWCNT were prepared by mechanical mixing during 1 minute and ultrasonication using UZDN-A (Russia) probe sonicator for 20 minutes with a frequency of 22 kHz and power of 130 W. During ultrasonic mixing the glass was constantly cooled in water. After cooling the hardener was added to the mixture that was used in the CFRP lay-up. Summary there were initial non-modified epoxy and 3 types of SWCNT-modified: with 0.1 wt.%, 0.2 wt.%, and 0.3 wt.% of SWCNT. In order to perform conductivity measurement of the CFRP specimens the tabs of silver conductive paint were prepared. The tabs were placed within a 30 mm distance and electrical conductivity was measured between these pads using LCR meter E7-20, Russia. There were 3 configurations for conductivity measurement: in-plane surface, out-of-plane diagonal, and out-of-plane through-thickness.

The specimens were tested using Instron 5582 universal testing machine. Quasi-static tension was performed at a rate of 1.5 mm/min with a gauge length of 70 mm. Three-point bending was conducted with a span of 60 mm at 10 mm/min rate. The initial structure and surfaces of fractured specimens were investigated by SEM using LEO EVO 50 (Zeiss, Germany) in "Nanotech" ISPMS SB RAS.

3. Experimental results and discussion

3.1. Electrical conductivity

The results of conductivity measurement are presented in Table 1. The in-plane conductivity of CFRP depends only on the surface fiber layer thus in-plane conductivity demonstrates the slight increase with the addition of CNTs however the CNT-0 specimen is conductive as well: 564 S/m. Quite identical results of out-of-plane through-thickness and diagonal measurements along with null conductivity for CNT-0 are showing that the final electrical conductivity of hybrid CFRP in out-of-plane mode is dictated by epoxy matrix. Thus the addition of CNTs enhances conductivity in final hybrid CFRP even if it was zero at the initial state without fillers. Table 1 shows the electrical properties for modified and non-modified specimens measured in three directions.

Table 1. Electrical conductivity of specimens.

Specimen	Electrical conductivity, S/m		
	CFRP in-plane surface	CFRP out-of-plane diagonal	CFRP out-of-plane through-thickness
CNT-0	564±319	$<10^{-5}$	$<10^{-5}$
CNT-1	514±252	6.8±1.5	7±1.18
CNT-2	702±228	16.1±1.5	16.3±1.02
CNT-3	865±184	65.3±11.1	66.2±9.9

3.2. Tensile and flexural testing

Three specimens were tested for each content of CNTs. The failure mode at the tension of all specimens was a lateral crack at the grip. Figure 2a demonstrates the plots of mean ultimate strength and modulus. It can be seen from the results in figure 2a that ultimate tensile strength and modulus are not significantly affected by addition of CNTs. The result is agreed with theoretical assumptions: tensile strength and modulus depend mainly on the longitude oriented fibers while the impact of a binder is small.

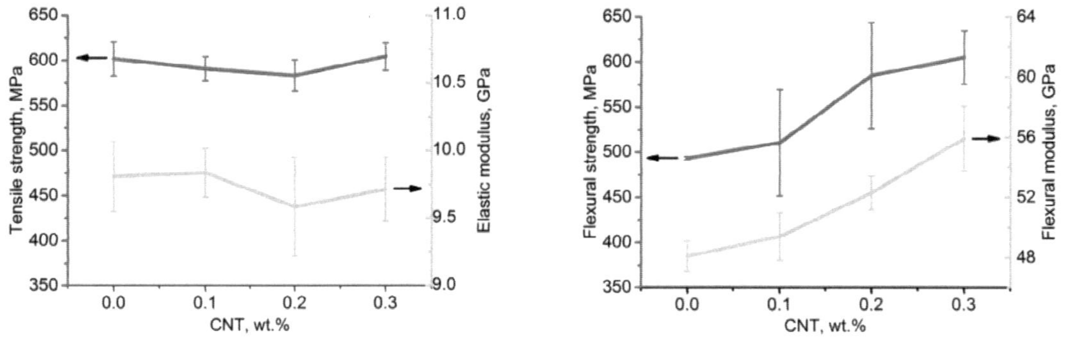

a b

Figure 2. Ultimate tensile strength and modulus (a) and flexural strength and modulus (b).

In flexural tests the strength and stiffness of the matrix greatly influence the final properties of CFRP/SWCNT hybrid composites. The shearing stresses occurring in interlayers of a bended specimen are born by the binder. Thus modification of matrix should improve flexural mechanical properties.

Figure 2b presents the results of a 3-point bending test. It can be seen that both flexural strength and modulus were increased by 23% and 16% correspondingly for the specimen with the addition of 0.3 wt.% of SWCNT. Such a modification increases the flexural strength of hybrid composite, but the scatter is quite high for CNT-1 and CNT-2. Flexural modulus with increase of CNTs demonstrates a uniform growth with the average scatter. The increase of flexural properties is attributed to higher shear strength of binder and adhesion between the layers after modification.

3.3. Scanning electron microscopy

Figures 3 and 4 show SEM images of initial structure and fractured epoxy specimens CNT-0 and CNT-3 respectively. The (a) images demonstrate the polished edge of the CFRP specimen while (b) and (c) are fractography images captured after flexural testing.

The figure 3a evidences the lack of conductivity resulting in the formation of the overexposed areas while for the specimen CNT-3 the image of the CFRP structure is quite uniform (figure 4a). As can be seen, both specimens have good uniformity of structure with no pores and damages.

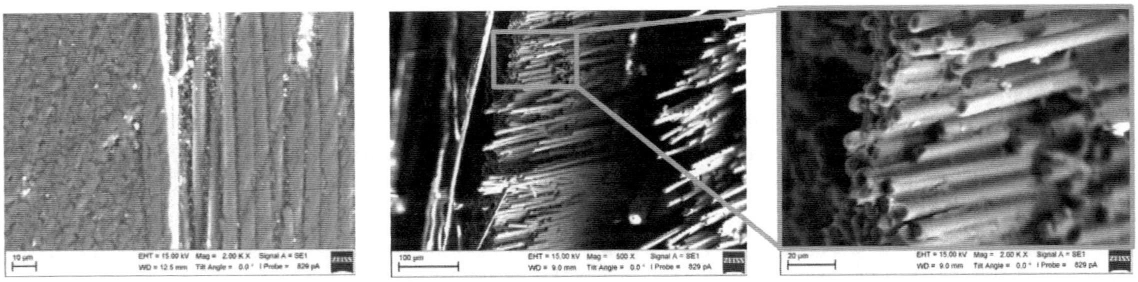

a b c

Figure 3. SEM images of CNT-0 specimen: initial structure (a) and fracture surface (b).

By comparison of fractography results of specimens CNT-0 and CNT-3 we can conclude that the fracture of these specimens occurs differently. Due to the increase of the elastic modulus and reduced viscoplastic properties of the modified binder, there form a lot of small crushed fragments between the filaments during testing of CNT-3 (figure 4c) while the non-modified epoxy binder remains good plasticity (figure 3c).

a b c

Figure 4. SEM images of CNT-3 specimen: initial structure (a) and fracture surface (b).

4. Conclusion

The hybrid CFRP/SWCNT composites were investigated. The research was focused on the evaluation of electrical and mechanical properties. It has been shown that tensile properties are less affected by the addition of SWCNT while flexural were improved. Flexural strength and modulus increase after the addition of 0.3 wt.% of SWCNTs by 22% and 16% correspondingly. Electrical conductivity after the addition of SWCNT increased as well. The effect of SWCNTs on CFRP conductivity is much higher for the out-of-plane path provided by interlayer connections in epoxy binder while in-plane conductivity determined by carbon fibers and therefore it shows an ambiguous increase with large scatter. The addition of 0.3 wt.% of SWCNT leads to increases of out-of-plane conductivity up to ~66 S/m while unmodified CFRP was dielectric. The method proposed in the work for producing hybrid composites has good potential and future research is to be linked to the development of reliable and inexpensive preparation and mixing technique and investigation of fatigue properties which are quite important for industrial application.

Acknowledgements
The research was supported by Grant of Russian Science Foundation №19-79-10148.

References

[1] Gibson R F 2010 A review of recent research on mechanics of multifunctional composite materials and structures *Compos. Struct.* **92** 2793–810

IOP Conf. Series: Materials Science and Engineering **731** (2020) 012002 doi:10.1088/1757-899X/731/1/012002

[2] Tehrani M, Safdari M, Boroujeni A Y, Razavi Z, Case S W, Dahmen K, Garmestani H and Al-Haik M S 2013 Hybrid carbon fiber/carbon nanotube composites for structural damping applications *Nanotechnology* **24** 155704

[3] Cortes L Q, Racagel S, Lonjon A, Dantras E and Lacabanne C 2016 Electrically conductive carbon fiber / PEKK / silver nanowires multifunctional composites *Compos. Sci. Technol.* **137** 159–66

[4] Jang J, Park H C, Lee H S, Khil M-S and Kim S Y 2018 Electrically and Thermally Conductive Carbon Fibre Fabric Reinforced Polymer Composites Based on Nanocarbons and an In-situ Polymerizable Cyclic Oligoester *Sci. Rep.* **8** 7659

[5] Zhao Z J, Zhang B Y, Du Y, Hei Y W, Yi X S, Shi F H and Xian G J 2017 MWCNT modified structure-conductive composite and its electromagnetic shielding behavior *Compos. Part B Eng.* **130** 21–7

[6] Panin S V., Duc Anh N, Kornienko L A, Alexenko V O, Buslovich D G and Ovechkin B B 2018 Wear-resistant polyetheretherketone composites *Mater. Today Proc.* **5** 25976–82

[7] Panin S V., Le T M H, Kornienko L A, Alexenko V O, Ivanova L R and Ovechkin B B 2018 Mechanical and tribotechnical properties of polyphenylene sulfide composites reinforced with carbon fibers of various dimension p 020231

[8] Caneba G T, Dutta C, Agrawal V and Rao M 2010 Novel Ultrasonic Dispersion of Carbon Nanotubes *J. Miner. Mater. Charact. Eng.* **09** 165–81

[9] Kim M and Kim J 2018 Nanoparticle dispersionizer by ultrasonic cavitation and streaming *Jpn. J. Appl. Phys.* **57** 07LE03

Algorithm for J-Integral Measurements by Digital Image Correlation

V V Titkov[1], A V Eremin[1,2], S V Panin[1,2], P S Lyubutin[1] and A A Kozulin[2,5]

[1]Institute of Strength Physics and Materials Science SB RAS, 2/4, pr. Akademicheskii, Tomsk, 634055, Russia
[2]National Research Tomsk Polytechnic University, 30, Lenin Avenue, Tomsk, 634050, Russia
[3]National Research Tomsk State University, 36, Lenina Avenue, Tomsk, 634050, Russia

svp@ispms.ru

Abstract. The work is devoted to the testing of the algorithm for calculating J-integral based on the construction of vector fields by digital image correlation (DIC) method. A comparative analysis of J-integral values calculated using DIC and instrumental data obtained in accordance with ASTM E 1820 "Standard Test Method for Measurement of Fracture Toughness" has made. It is shown that this approach can be used for cases when the standard technique for measuring the J-integral cannot be applied, or the standard technique does not allow achieving the required accuracy for the integral determination in local areas of the loaded material.

1. Introduction

The digital image correlation (DIC) method [1, 2] is becoming increasingly popular due to its versatility, ease of use, and the ability to obtain full-sized displacement/deformation fields when processing images of the surface of a loaded object. Based on the calculated fields, it becomes possible to determine the required parameters of the stress-strain state, for example, such as SIF, CTOD, J-integral and other characteristics adopted in fracture mechanics.

J-integral proposed by Cherepanov and Rice in 1967-68 [3, 4] is defined as a contour integral independent of the path which characterizes stresses and strains in the region of the crack tip in elastic or elastoplastic loaded materials. The main questions arising in determining the J-integral using direct integration over the contour that шincludes the crack tip with a gap on its edges are the choice of the integration contour, in particular, its shape, dimensions, position relative to the crack tip, and the effect of these parameters on the accuracy of final calculations. According to the definition of the J-integral it is independent on the integration path and this property is satisfied when working with model data [5]. However, in an experiment, due to errors in determining the displacements resulting from the use of the digital image correlation method, independence from the integration path is not fulfilled [6]. In the presented study, an algorithm for measuring the J-integral is described and tested with a help of data obtained using the digital image correlation method, and they are compared with the results obtained by the instrumental method (by means of COD gauge).

2. Algorithm for j-integral measurement

The process of measuring the J-integral (including the allocation of the plastic zone) when using the approach to calculating the optical flux during mechanical testing of materials can be divided into four stages:

1. calculation of displacement fields [7];
2. post-processing of the fields [8];
3. calculation of the components of strain and stresses [5];
4. calculation of the J-integral.

Since research on stages 1, 2 and 3 has published in the literature and describes the algorithms in sufficient detail, the results devoted only to aspects of the calculation of the J-integral are given below.

2.1. Calculation of the J-integral [5]

For known values of displacements, strains, and stresses, the J-integral [9] can be calculated by the well-known formula [3]:

$$J = \int_\Gamma \left(w\,dy - T\frac{\partial u}{\partial x}ds\right), \tag{1}$$

where w – strain energy density:

$$w = \int_0^\varepsilon \sigma\, d\varepsilon,$$

where D – plane stress stiffness matrix, E – elastic modulus, ν – Poisson's ratio. Thus, when the fields of the stress and strain tensor are known the J-integral can be estimated from the equation (1) as follows [10]:

$$J = \int_\Gamma \frac{1}{2}\left(\sigma_{xx}\varepsilon_{xx} + \sigma_{yy}\varepsilon_{yy} + 2\tau_{xy}\gamma_{xy}\right)dy - \left[\sigma_{xx}n_x + \tau_{xy}n_y, \tau_{xy}n_x + \sigma_{yy}n_y\right]\cdot\begin{bmatrix}\frac{\partial u}{\partial x}\\ \frac{\partial v}{\partial x}\end{bmatrix} ds, \tag{2}$$

Γ – a contour covering the crack tip and traversing counter-clockwise; n_x, n_y – the components of the vector perpendicular to Γ; u, v – components of the displacement vector.

The contour used to calculate the J-integral should not only cover the crack tip but also lie outside the plastic zone. The shape of the plastic region was determined using a two-dimensional Mises which criterion for plane stress conditions as follows:

$$\sigma_{Mises} = \sqrt{\sigma_x^2 + \sigma_y^2 - \sigma_x\sigma_y + 3\tau^2},$$

where σ_{Mesis} – stress at a certain point on the surface according to Mises criterion, σ_x – normal stress in the *x*-direction, σ_y – normal stress in the *y*-direction, τ – shear stress (σ_{xy}). When the Mises criterion is higher than the yield stress (σ_{yield}), it is assumed that the plastic deformation occurs:

$$\sigma_{Mises} \geq \sigma_{yield}$$

Figure 1 (c) shows the visualization of the plastic zone (according to the Mises) with excluded areas of the grips and the notch.

3. Algorithm testing and results

To test the proposed method for calculating the J-integral the obtained results were compared with the data from Table 1 calculated on instrumental measurements according to the standard [11]. The optical method and the J-integral calculation method according to the standard were applied to the results of uniaxial tension of a compact specimen (10 mm thick with a 21.8 mm long fatigue crack, figure 1, a) from AA2024 aluminum alloy (figure 1, b) (the elastic modulus of the alloy is 71 000 MPa, Poisson's ratio 0.33, yield strength 286.7 MPa).

3.1. Estimation of the J-integral deviation

To obtain a quantitative assessment of the accuracy and noise immunity of the J-integral calculation, it is proposed to use the following value:

$$D = \left|\frac{J_{ASTM} - J_{DIC}}{J_{ASTM\,max}}\right| * 100\%,$$

where J_{ASTM} –J- integral obtained according to ASTM standard; J_{DIC} – the value of the J-integral determined using DIC, $J_{ASTM\,max}$ – the maximum value of the J-integral obtained according to [11].

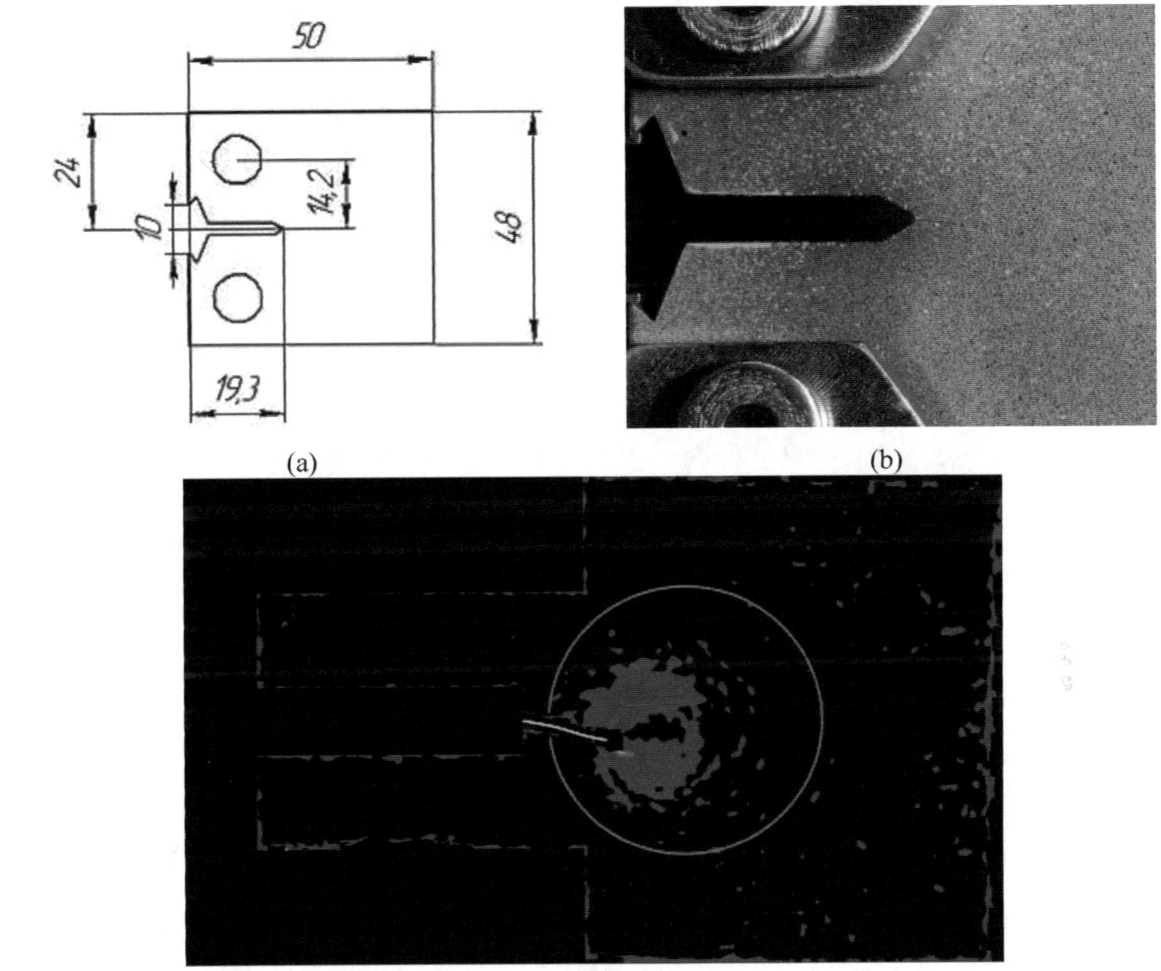

Figure 1. (a) The dimensions of a compact specimen with a thickness of 10 mm and fatigue crack of 21.8 mm long; (b) The original photo of the specimen with the speckle pattern on the surface for DIC evaluation; (c) An appearance of the plastic zone (according to the Mises) near the crack tip. The red line denotes integration contour. Crack wake is shown by the green line.

According to the standard, the values of the J-integral were calculated (Table 1, figure 2). The deviation between the J-integral value calculated using the DIC method relative to that determined one found using the COD gauge vary from 0.695 % to 4.088 %. Moreover, the average error for the analyzed range of load is 1.753 %.

Table 1. Values of the J-integral and their deviations from the instrumentally measured ones.

Value	Load, kN										
	1.102	1.906	2.704	3.649	4.304	5.1	5.9	6.701	7.499	8.301	9.096
J_{ASTM}, N/m	208	624	1256	2303	3184	4471	5983	7675	9815	12736	17185
J_{DIC}, N/m	328	953	1500	2214	2844	3865	5281	7205	9742	13032	17139
D, %	0.695	1.911	1.418	0.519	1.975	3.526	4.088	2.735	0.421	1.719	0.268

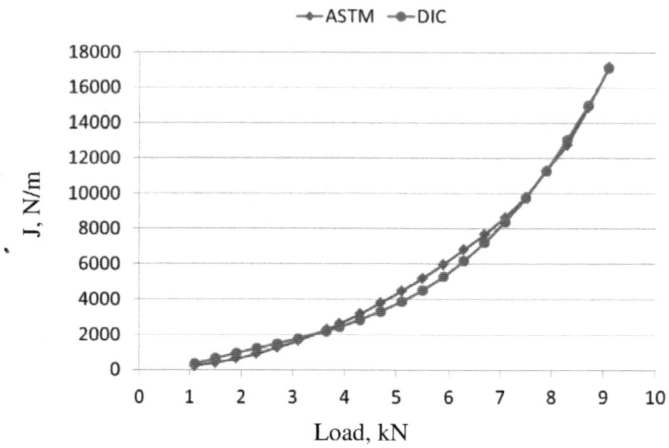

Figure 2. The change in the value of the J-integral with increasing applied load: (1) calculated according to ASTM standard, and (2) according to the proposed method.

4. Conclusion

Application of the J-integral measurement algorithm using the DIC method which consists in constructing a field of displacement vectors, calculating deformations and stresses, and then calculating the value of the specified parameter of fracture mechanics has been considered. As a result of testing the algorithm it was found that the deviation of the calculated values of the J-integral from the values instrumentally measured using the COD gauge (according to ASTM) averages 1.753 % at a minimum and maximum deviation 0.695 % and 4.088%, respectively. Thus, this approach can be used in cases when the standard technique for measuring the J-integral cannot be applied, or the integral nature of the measurements does not allow to achieve the required accuracy for determining the loaded material in local areas.

Acknowledgement

The work was performed in the framework of fundamental research Program of the Russian State Academies of Sciences for 2013–2020, line of research III.23. RF President Council Grants i) for the support of young researchers No. MK-6762.2018.9, ii) leading research schools No. NSh-5875.2018.8 is acknowledged. Authors also appreciate partial support of RFBR Grants No. 18-47-700006.

References

[1] Sutton M A, McNeill S R, Helm J D and Chao Y J 2000 *Photomechanics* pp 323–72
[2] Sutton M A, Orteu J J and Schreier H 2009 *Image Correlation for Shape, Motion and Deformation Measurements: Basic Concepts, Theory and Applications* (Springer Science & Business Media)
[3] Rice J R 1968 *J. Appl. Mech.* **35** 379–86
[4] Becker T, Mostafavi M, Tait R and Marrow T 2012 *Fatigue Fract. Eng. Mater. Struct.* **35** 971–84
[5] Titkov V V, Panin S V, Eremin A V, Kozulin A A and Lyubutin P S 2018 *AIP Conf. Proc.* **2051** 020305

IOP Conf. Series: Materials Science and Engineering **731** (2020) 012003 doi:10.1088/1757-899X/731/1/012003

[6] Tracy J, Waas A and Daly S 2015 *J. of Mat. Sc.* **50**(13) 4646–58
[7] Panin S V, Titkov V V and Lyubutin P S 2014 *Avtometriya* **50**(2) 39–49
[8] Panin S V, Titkov V V and Lyubutin P S 2014 *Opt., Instr. and Data Proc.* **50**(1) 61–7
[9] Cherepanov G P 1967 *J. Appl. Math. Mech.* **31**(3) 503–12
[10] Vavrik D and Jandejsek I 2014 *Eng. Fract. Mech.* **129** 14–25
[11] ASTM E 1820 Standard Test Method for Measurement of Fracture Toughness (ASTM International, 2014)

IOP Conf. Series: Materials Science and Engineering **731** (2020) 012004 doi:10.1088/1757-899X/731/1/012004

Deformation Behavior under Static and Cyclic Tension of Polymer Grafts without and after Modification by RGD Peptides

S V Panin[1], L V Antonova[2], A V Byakov[1], T V Glushkova[2], E O Krivkina[2], A V Eremin[1], V N Silnikov[3] and L S Barbarash[2]

[1]Institute of Strength Physics and Materials Science SB RAS, 2/4, pr. Akademicheskii, Tomsk, Russia

[2]Research Institute for Complex Problems of Cardiovascular Diseases, Kemerovo, Russian Federation

[3]Institute of Chemical Niology and Fundamental Medicine SB RAS, Novosibirsk, Russian Federation

svp@ispms.ru, antonova.la@mail.ru

Abstract. The structure, mechanical properties, and deformation behavior under static and cyclic tension of biofunctionalized biodegradable vascular grafts based on polyhydroxybutyrate/valerate and polycaprolactone were studied. It is shown that the modification gives rise to an almost twofold decrease of the elongation at break as well as the ultimate strength. It is shown that under cyclic loading the modification of grafts results in decreasing cyclic durability by more than twice. In doing so, the level of deforming stress decreases to a much lesser extent and is practically inferior to that for unmodified material. The analysis of principal strain ε_1 and ε_2 component distribution patterns in grafts of both types is carried out while the reason for the observed changes is discussed.

1. Introduction

To date, clinical practice still lacks ready-to-use vascular prostheses (grafts) of small diameter suitable for in situ regeneration of blood vessels. They are of actuality for carrying out shunting operations with replacing damaged vessels. However, the use of synthetic non-biodegradable polymers for manufacturing vascular grafts of small diameter was not successful due to development of chronic inflammation, thrombosis and neointima hyperplasia [1, 2].

For manufacturing vascular prostheses in situ many natural and artificial biodegradable polymers are used [3]. Natural polymers possess high biocompatibility and in vivo are completely subjected to gradual degradation to non-toxic substances. However, they give rise to a pronounced immune response; also, they have poor manufacturability while their mechanical properties are far distinct from blood vessels [4]. On the other hand, synthetic polymers do not usually cause a strong immune response; possess appropriate mechanical characteristics and high processability while their disadvantage is low biocompatibility: the absence of sites which promote cell adhesion [3].

Recently, a number of strategies have been developed to provide polymer surfaces the ability to selectively adhere endothelial cells in order to accelerate endothelization. Most of them are based on immobilization or fixation on the surface of specific cell adhesion proteins and bioactive peptides [3].

Arginine-glycine-asparagine (RGD) is a cell adhesion site that presents in many extracellular matrix proteins [5]. This sequence is one of the key ligands for integrins — receptors that are responsible for cell adhesion, migration, proliferation, differentiation, and survival [6]. The RGD peptide can be considered as a common integrin binding motif. Compared to proteins, bioactive RGD-containing peptides possess a simpler structure and higher chemical stability. Due to artificial synthesis, it is possible to develop a variety of configurations of RGD peptides with low immunogenicity and different potential for interaction with cells.

Biofunctionalization of synthetic materials allows one to create biomimetic surfaces capable of generating specific cellular responses. The incorporation of biologically active peptides into the tubular scaffolds during the electrospinning process in combination with providing the created structure with functional activity can change the architectonics of surface. The latter should affect their deformation behavior under cyclic loading. The aim of the work was to study the deformation behavior under static and cyclic loading of biofunctionalized biodegradable vascular grafts based on polyhydroxybutyrate/valerate and polycaprolactone without and after modification by RGD peptides.

2. Materials and research methods

Tubular polymer matrices with a diameter 1.5 mm were fabricated by electrospinning with the use of a "Nanon-01A" device (MECC Co. Ltd, Japan). The composition of polyhydroxybutyrate/valerate (PHBV) polymers with 10 % valerate (Sigma-Aldrich, USA) and polycaprolactone (PCL) with a molecular weight of 80,000 kDa (Sigma-Aldrich, USA) were employed. Nonpolar organic chloroform ChP (Vecton, Russia) was used as a solvent. The ratio of polymers in solution made 1:2.

Hexamethylene diamine, glutaraldehyde, ninhydrin, ascorbic acid (Sigma-Aldrich, USA) and RGD (NanoTech-S, Russia) were used for surface modification of the polymer tubular frameworks made of the PHBV/PCL with the RGD peptides. First, the frameworks were treated with a 10 % solution of hexamethylenediamine in a water solution of isopropanol (1:1 ratio). The treatment was conducted at 37 °C for 30 min which reduced the modification time twice [7, 8]. For the further treatments, each graft was put horizontally in a separate reactor (with a volume of 15 ml) with constant swaying.

Static tension tests were carried out with the use of a Zwick/Roell universal testing machine (Zwick GmbH, Germany). The force gauge of a nominal 50 N capacity was employed with cross head velocity 10 mm/min. The tensile strength was estimated as the maximum tensile stress (MPa) before the onset of fracture. Elastic properties were estimated by the relative elongation at break (%) and Young's modulus (MPa).

The structure of the inner grafts' surface as well as wall thickness of non-woven tubular PHBV/PCL frameworks (at observation area of 0.5 × 0.5 mm) without and after the RGD peptides modified was studied using S-3400N scanning electron microscope (Hitachi, Japan) under high vacuum. For the observation, specimens were sprayed with gold and palladium foils with the help of a SC-7640 vacuum unit (Quorum Technologies, England).

The Digital Image Correlation technique (DIC) was used to analyze the deformation behavior of loaded vascular grafts. The VIC 3D system (Correlated Solutions) was employed. Specimens in the initial and modified state were loaded at static tension with the use of an Instron 5582 electromechanical testing machine. For this, fragments 60 mm long were cut out from tubular blanks. Then fine paint particles were sprayed with the use of an airbrush to form a contrast speckle pattern. The flattened (initially tubular shaped) specimens after clamping in grips had a strip-like shape. This is motivated by the fact of fiber damaging being cut along the axis. For this reason they became extremely heterogeneous and their elongation at failure was noticeably reduced. Then, such specimens were clamped into wedge grips and stretched at the loading rate 10 mm/min. Surface images were recorded every 10 seconds with the help of Canon EOS 700 digital SLR camera equipped with a telescopic lens. Further, using the DIC technique strain fields were calculated. Then dependencies of the average principal strains ε_1 and ε_2 versus tensile strain were constructed.

To study the deformation behavior of specimens under cyclic tension a tubular graft was cut along the axis. Dog-bone shape specimen had the dimension shown in figure 1a. One of the folds oriented along the specimen axis remained in the center (Figure 1b). However the latter did not affect their deformation behavior. A speckle pattern consisted of black alkyd paint spots over white background and was deposited to the outside surface of the specimen (Figure 1b).

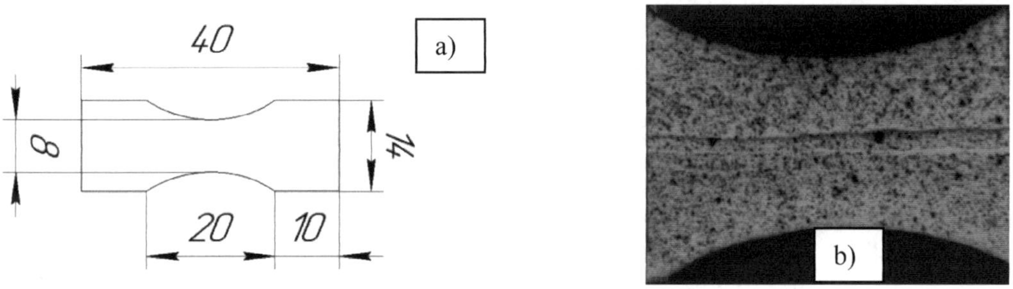

Figure 1. Dog-bone shape specimen sketch (a) and speckle pattern photograph (b) of a vascular graft.

The tests were carried out with the use of a Makron servo-hydraulic machine (Biss, India). The maximum cross head displacement is 60 mm with the accuracy of 1 μm. Specimens were tested by blocks of 100 cycles with the control over the grip movement (control by deformation). The amplitude of the cross head displacement was equal to 2 mm. After each cyclic loading block the surface was imaged. Then, the cross head was displaced to compensate the specimen loosening that took place during the cyclic loading. The average elongation of the specimens during 1 block of cycling was 0.25 mm, i.e. for every 100 cycles the specimen was extended by ~ 1.25 %. Canon EOS 700D camera with an EF 70-300 mm telephoto lens was used. The necessary zoom was ensured by two macrolenses (+1, +2 diopters).

3. Results and discussion

Modification density of the prostheses' surface with the RGD peptides at the treatment for 30 minutes was 6.3 nM/cm^2. The presence of peptides on the polymer surface was confirmed with the use of the Sakaguchi test over arginine presence [7, 8]. The specimen was remained colored after their washing that indicated covalent bonding of the peptide. A control specimen containing no RGD peptide stained light yellow; its coloring removed upon washing. Thus, the efficiency of the RGD peptide modification of the PHBV/PCL tubular framework surfaces at the modification time 30 minutes was confirmed.

The PHBV/PCL grafts in both states possess a highly porous structure with a wall thickness 170–250 μm consisting of homogeneous fibers (Figure 2). The average fiber diameter and average pore area in the grafts were 2.63 ± 1.14 μm and 47.13 ± 23.0 μm^2, respectively, with the average porosity of ~50.0 %.

Stiffness is a particularly crucial indicator for the subsequent vessel compliance into which the prosthesis is planned to be implanted. Significant discrepancies in these parameters of the prosthetic vessel and vascular one may subsequently provoke neointima hypertrophy or thrombosis. Therefore, the stiffness diminishing was one of the main missions at PHBV/PCL based vascular grafts modification by the RGD peptides. It is shown that modification of the PHBV/PCL grafts with the RGD peptides gives rise to variation of their physic-mechanical properties (Table 1): a 1.7-fold decrease in elongation at break takes place against a 1.6-fold decrease in stiffness in regard to the unmodified analogues ($p < 0.05$).

Table 1. Physico-mechanical properties of biodegradable vascular grafts without and after RGD modification

Specimen	Tensile strength, MPa	Young's modulus, MPa	Elongation at break, %
PHBV/PCL	7.1	28.0	431.3
PHBV/PCL + RGD	1.8	17.8	248.9

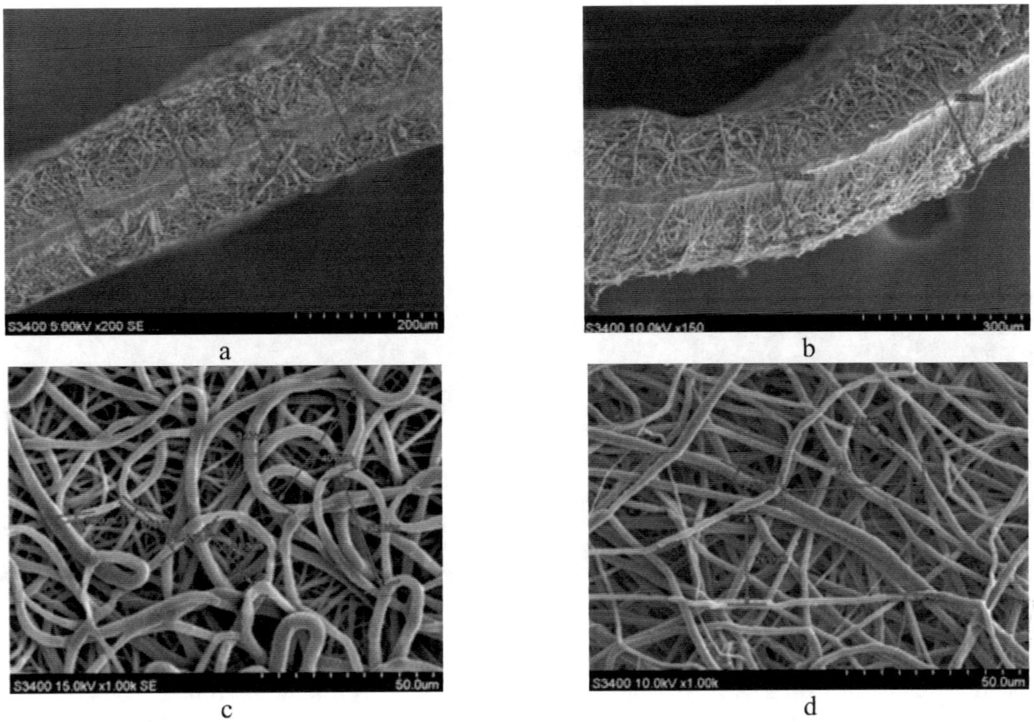

Figure 2. SEM micrographs of PHBV/PCL grafts without and after modification with the RGD peptides: cross section of the wall of the unmodified PHBV/PCL (a) ×200 and PHBV/PCL + RGD (b), ×150; the inner surface of the unmodified PHBV/PCL (c), ×1000, PHBV/PCL + RGD (d), ×1000

3.1. Mechanical behavior of biodegradable grafts under static loading

Figure 3 shows the dependences of the principal strains ε_1 and ε_2 versus tensile strain for the grafts in both states. It is seen that, for the above reason, the elongation at break was reduced less than 2 times while the lowering patter was kept. The graph of the principal strain component ε_1 looks less uniforms; in the case of the nonmodified graft while the inclination angle is slightly lower (Figure 3a). However, the graphs of another component ε_2 differ significantly (Figure 3b). Thus, this deformation parameter turned out to be very sensitive to the structural changes happened. To understand the observed differences the maps of the deformation component distribution were then analyzed.

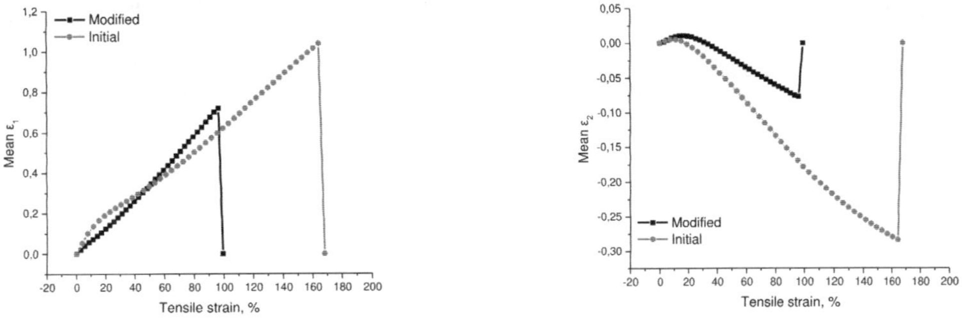

a) – ε_1 b) – ε_2

Figure 3. Diagrams of the principal stain ε_1 (a) and ε_2 (b) versus tensile strain at static tension of vascular grafts specimens without (initial) and after the modification (modified)

IOP Conf. Series: Materials Science and Engineering **731** (2020) 012004 doi:10.1088/1757-899X/731/1/012004

Maps of the principal strains ε_1 (a) and ε_2 distribution of the initial state and modified grafts at relative strain 20 % are given in figure 4. It is seen that the onset of plastic deformation is not accompanied by any pronounced strain localization (not taking into account the regions in the middle specimen where some extreme values are evident; however, they are related to the shrinkage of the half folded tube). The pattern of ε_2 component distribution shows in the modified graft (which possesses less ductility) exhibits less uniformity (Figure 4d). By the authors opinion the latter is due to the heterogeneous structure of the modified graft.

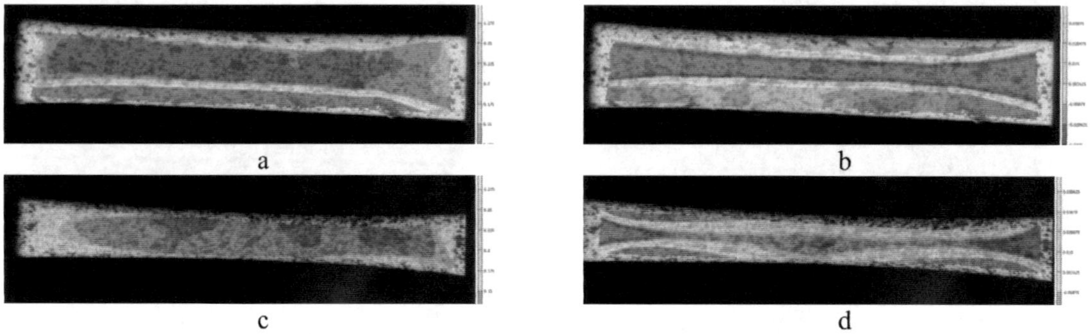

Figure 4. Maps of ε_1 and ε_2 distribution of vascular grafts without and after the modification (c, d); static stretching; strain level 20 %; a), c) ε_1; b), d) ε_2; a), b) without modification; c), d) after modification

With further loading the specimens (in both states) are tightened; in doing so the strain distribution is fairly uniform in the initially half-folded tube specimens. To a greater extent this is related to the ε_1 component distribution. However, another principal strain ε_2 (Figure 5b,d) looks less uniform. Note, that in the modified grafts the average strain level is higher: in the middle part of the specimen it exceeds that for the nonmodified graft up to 2 times.

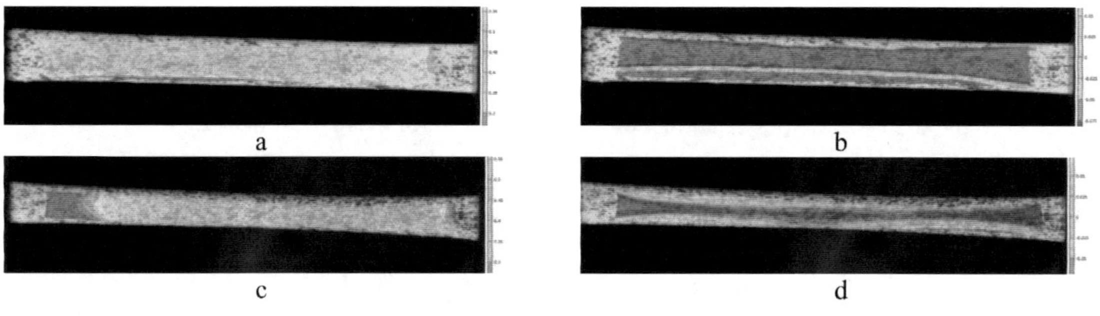

Figure 5. Maps of ε_1 and ε_2 distribution of vascular grafts without and after the modification (c, d); static stretching; strain level 80 %; a), c) ε_1; b), d) ε_2; a), b) without modification; c), d) after modification

Generally, a similar deformation pattern is kept up to critical strains preceding specimens' fracture (Figure 6). Despite high straining (the elongation at break exceeds 100%) bothe types of specimens fail without neck formation or any other form of strain macrolocalization. This is due to both i) tubular shape of the specimens and ii) fibrous structure of the material.

3.2. Mechanical behavior of biodegradable grafts under cyclic loading

The cyclic tension test data evidences for the nonmodied graft failure after 82 blocks of loading (100 cycles each). In so doing, the length of the specimen has enlarged by 202.5 %. The RGD peptides modified graft has failed after applying 33 cyclic loading blocks with relative elongation (at break) 142.5 %. Thus, in contrast to the data of static tension tests (Table 1), the difference in the relative elongation has reduced from 1.73 down to 1.42 times.

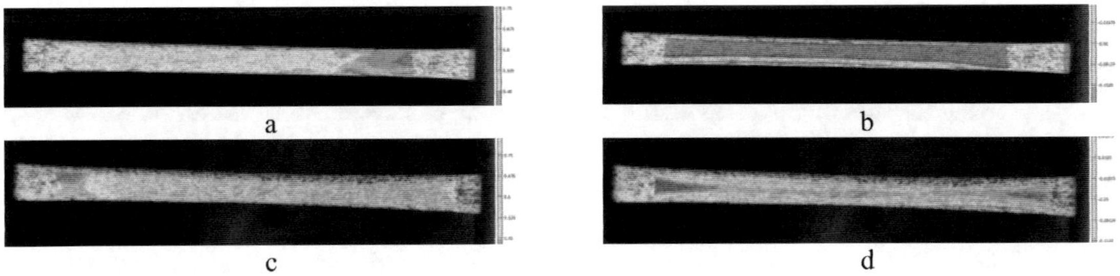

Figure 6. Maps of ε_1 and ε_2 distribution of vascular grafts without and after the modification (c, d); static stretching; a), c) ε_1; b), d) ε_2; a), b) without modification; c), d) after modification; a), b) without modification (deformation degree – 140 %); c), d) after modification (deformation degree – 96 %)

On the other hand, under cyclic tension (at the displacement control loading), the magnitude of the applied load differs insignificantly. This means that regardless less cyclic resistance under used severe testing conditions they possess nearly equal baring capacity. In combination with the fact that mechanical properties of the modified material are closer to native vessels it brings promise of their wide practical applications.

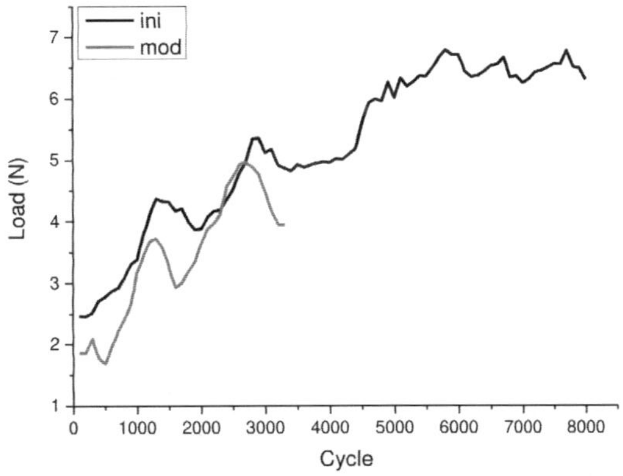

Figure 7. Graph of the applied load versus the number of loading cycles for the nonmodified (ini) and modified (mod) graft specimens

Figure 8 shows patterns of principal strain ε1 distribution with increasing cyclic loading. It is seen that in the nonmodifed graft no signs of strain localization are revealed almost 300 cycles prior to failure N = 8·103 (Figure 8 a,c,e,g). Only at "necking" tearing takes place with cracks arising from opposite sides of the dog-bone shaped specimen to give rise to the main crack propagation (Figure 8i).

In the modified graft a strain macrolocalization is already fixed at the low cycle number $N = 2\cdot10^3$. Soon, the tearing started being initiated from one of the edges (Figure 8f). In this case, layered pattern of specimen fracture is observed when the upper layer begins to break while the underlying layers maintain continuity (Figure 8h). Just after, the specimen withstands another 300 loading cycles and fails (Figure 8j). The reason is obviously related to its heterogeneity induced by the structure modification.

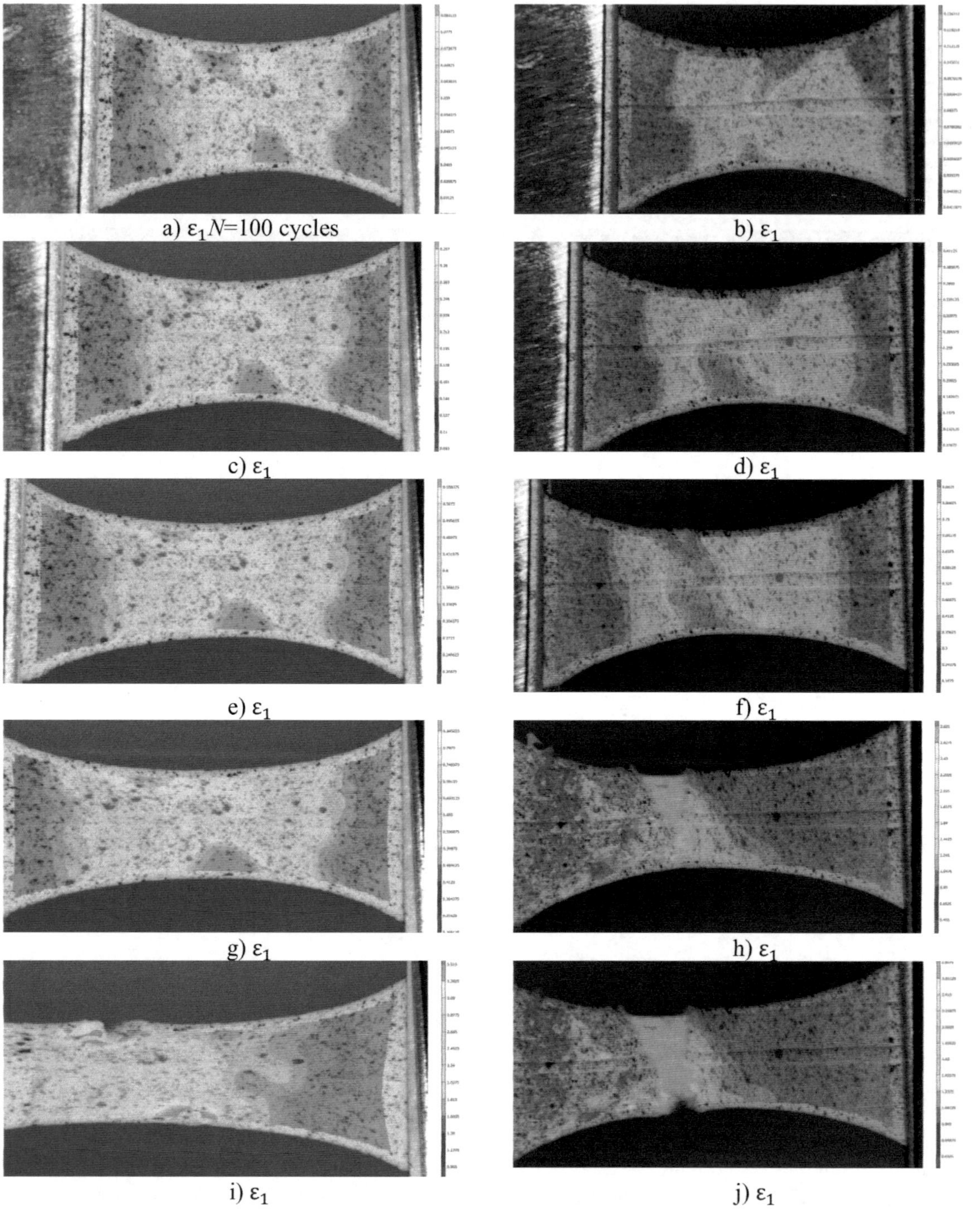

Figure 8. Principle strain ε_1 maps for the vascular graft without and after modification; cyclic stretching; The number of cycles: a, b) 100, c, d) 1000, e, f) 2000, g, h) 3000, i) 8000, j) 3300; a, c, d, g, i) without modification; b, d, f, h, j) after modification

4. Conclusion

The structure, mechanical properties, and deformation behavior of polymer grafts in the initial and modified state were studied. The used modification regime of polymer biodegradable tubular PHBV/PCL grafts with RGD peptides (during 30 minutes) did not result in a statistically significant variation of their surface structure. Moreover, after the modification, the stiffness of grafts and their elongation ability decreased. This has brought the physico-mechanical properties of vascular prostheses PHBV/PCL + RGD closer to ones of native vessels.

Experimental studies on the deformation behavior of biodegradable grafts under static and cyclic tension was carried out. The analysis and visualization of strain fields by means of digital image correlation was conducted. It is shown that principal strain distribution patters correlates well with variation of mechanical properties, namely, the plastic flow stress and elongation at break. It is shown that principal strain component $\varepsilon 2$ in the modified graft (possessing lower mechanical properties is distributed less uniformly while its level prior to failure is up to two times higher than that in the unmodified material.

Regardless less cyclic resistance under used severe testing conditions the modified grafts possess nearly the equal baring capacity under cyclic tension. In combination with the fact that their mechanical properties are closer to native vessels it brings promise of their wide practical applications.

Acknowledgement

The research was funded by the integration project "Biodegradable Vascular Graft of Small Diameter Modified with Functionally active peptides"of the complex program of fundamental scientific research of the SB RAS II.1 (№ 0546-2018-0001). The part of the project "Evaluation of mechanical properties of initial and modified grafts under static and alternating loading. Multilevel simulation of deformation behavior of initial and modified grafts" was carried out in ISPMS SB RAS. The part of the project "Development of scientific fundamentals for the manufacture and modification of a biodegradable vascular graft and its *in vitro* testing" was conducted at the RI CPCD.

References

[1] Tatterton M, Wilshaw S P, Ingham E Homer-Vanniasinkam S. 2012 *Vascular and Endovascular Surgery* **46** (3) pp 212–222
[2] Desai M, Seifalian A M, Hamilton G 2011 *European Journal of Cardio-Thoracic Surgery* **40**(2) pp 394–398
[3] Ren X, Feng Y, Guo J, Wang H, Li Q, Yang J, Hao X, Lv J, Ma N, Li W. 2015 *Chem. Soc. Rev.* **44** (15) pp 5680-5742
[4] Ingavle G C, Leach J K 2014 *Tissue Engineering. Part B: Reviews* **20**(4) pp 277–293
[5] Wang F., Li Y., Shen Y., Wang A., Wang S., Xie T. 2013 *Int J Mol Sci* **14**(7) pp 13447–13462.
[6] Antonova L V, Silnikov V N, Sevostyanova VV, Yuzhalin A E, Koroleva L S, Velikanova E A, Mironov A V., Godovikova TS, Kutikhin AG, Glushkova T V, Serpokrylova IYu, Senokosova E A, Matveeva V G, Khanova M Yu, Akentyeva T N, Krivkina EO, Kudryavtseva Yu A, Barbarash LS 2019 *Polymers* **11**(174) pp 1-18
[7] Lin H B, Sun W, Mosher D F, García-Echeverría C, Schaufelberger K, Lelkes PI, Cooper S L 1994 *J Biomed Mater Res* **28**(3) pp 329–342
[8] Parniak M A, Lange G, Viswanatha T 1983 *J Biochem Biophys Methods* **7**(4) pp 267–276

IOP Conf. Series: Materials Science and Engineering **731** (2020) 012005 doi:10.1088/1757-899X/731/1/012005

Application of Laser Welding in Car Bodies Manufacturing

J Viňáš[1], J Brezinová[1], J Brezina1, P Maruschak[2] and S Panin[3]

[1]Technical University of Košice, Mäsiarska 74, 040 01 Košice, Slovakia
[2]Ternopil National Ivan Pul'uj Technical University, Rus'ka str. 56, 46001 Ternopil, Ukraine
[3]Institute of Strength Physics and Materials Science of Siberian Branch of Russian Academy of Sciences, 634055 Tomsk, Russia

svp@ispms.ru

Abstract. The paper deals with a problem of laser welding of zinc-coated steel (DX53D+Z and DX54D+Z) plates. Experiments were carried out with the use of steel plates 0.8; 1.0; 1.3; and 1.75 mm thick. High power 2.5-2.9 kW CO_2 laser was employed. CO_2 was used as the shielded gas. The optimal parameters of welding for material under study were determined. The mechanical properties of welded joints were measured through the tensile loading and microhardness measurement. Tensile performances of the specimens fabricated at different welding rate and laser beam power were compared with that of the base material.

1. Introduction

Combined tailored welded blanks (TWB) consist of two or more joined pieces and they can represent a combination of materials of different thickness and grades. These tailored welded blanks found their application mostly in automobile industry where reliably is of crucial importance. The joining technique allows one to influence hardness, stiffness or corrosive resistance of metal sheets in different parts of the component. In doing so, the hardness might be increased, while parts of different thickness might be connected even being covered with protective layers (foils) [1-4].

It is known that zinc melts at a temperature of 906 ° C, while steel at 1100 ° C. With the conventional method of welding, the coating is damaged inevitably. During laser welding of galvanized steel sheets, a high-pressure evaporation region (66–1060 atm) arises around the melt with a temperature of 1600–3000 °C [5-7]. If the gap between the sheets is narrow, the vapor cannot go away, deviations occur in the pressure of the vapor-gas channel, and damage takes place. An explosive degassing process disrupts the welding process, and the vapor-gas channel is destroyed due to pressure imbalance. A large volume of molten steel is forced out of the molten pool, which leads to the formation of shells, imperfections or pores in the weld metal, which affect the mechanical properties of the weld [8–10]. The output and the laser welding rate were taken as fundamental parameters affecting the quality of the welds. Note, that the output is the most critical factor in the laser welding operation [11-15].

The aim of the study is to evaluate the application of laser welding technology for joining the materials of specified thickness. Improper external and internal welding parameters were found using non–destructive and destructive (mechanical) tests. Also, some reasons for the structure heterogeneity nucleation in the welds and their influence on mechanical properties were revealed. In doing so, standard and small test coupons were examined at two loading rates. The next parts of the paper deals with the study of Zn surface layer behaviour of the combined tailored blanks during the welding process. To this end, the microhardness of welds was measured. The possibility of the formation of martensite phase is shown and discussed.

2. Experimental part

The material under study was thermally zinc-coated with steel sheets (10.0 – 12.5 μm thick) used in the automobile industry [1]. Samples were welded by CO_2 laser ELB5. The following welding parameters were used: TWB1 = 2.5 kW, TWB2 = 2.9 kW; welding rate was 45 mm /s^{-1}, the pressure of protective gas Ar was 0.2 MPa, beam was focused at the surface. Welds were non–destructively tested by visual examination (according to EN ISO 17637:2016), as well as Penetrant inspection (PT) according to EN ISO 23277:2010 for the detection of surface faults. Radiographic testing was employed to detect internal faults (EN 12517-1:2006-10). Chemical composition and mechanical properties (yield strength - σ_{ys}; tensile strength – σ_{ts}; elongation at break – ε_t [%]) of tested materials i) DX53D+Z and ii) DX54D+Z EN 10346: 2015 are shown in Table 1 and Table 2.

Table 1. TWB1 – materials combination.

	Thickness	**Chemical composition of TWB 1 (wt. %), Fe - bal.**							
Material	(mm)	C_{max}	Mn_{max}	P_{max}	S_{max}	Si_{max}	Al	Ti	N_{max}
DX53D+Z	1	0.04	0.20	0.015	0.012	0.01	0.03-0.06		0.006
DX54D+Z	0.8	0.015	0.20	0.015	0.015		min 0.02	0.06-0.14	0.006

	Mechanical properties[1)]			
Material	σ_{ys} (MPa)	σ_{ts} (MPa)	ε_t (%)	Bending angle 180°
DX53D+Z	260	max. 380	30	0.a
DX54D+Z	220	max. 350	36	0.a

Table 2. TWB2 – materials combination.

	Thickness	**Chemical composition of TWB2 (wt. %), Fe - bal.**								
Material	(mm)	C_{max}	Mn_{max}	P_{max}	S_{max}	Si_{max}	Al	Ti	Nb_{max}	N_{max}
DX53D+Z	1.75	0.04	0.20	0.015	0.012	0.01	0.03-0.06			0.006
DX54D+Z	1.3	0.01	0.6	0.025	0.08	0.04	0.15	0.04	0.003	

	Mechanical properties [1)]			
Material	σ_{ys} (MPa)	σ_{ts} (MPa)	ε_t (%)	Bending angle 180°
DX53D+Z	260	max. 380	30	0.a
DX54D+Z	260	max. 350	24	0.a

[1)]Tests were done in the transversal direction in regard to rolling direction

Then, mechanical testing of the specimens was carried out. Tensile test (according to EN ISO 4136:2011-09) were performed for the samples with a transverse welds of standard and reduced dimension. Also, the different loading rate was employed. For the standard (S) (l = 80 mm) (figure 1) and non – standard (*N*) (l = 30 mm) (figure 2) samples, the loading rate was v=30 mm . min^{-1} and v=0.9 mm . min^{-1} respectively. Metallographic analysis and microhardness measurement were conducted according to EN ISO 9015-2:2011-09 (HV0.01).

The EDX analysis over the specimens' cross section was conducted with further microhardness measurement in the WM (weld metal) and heat affected zone (HAZ).

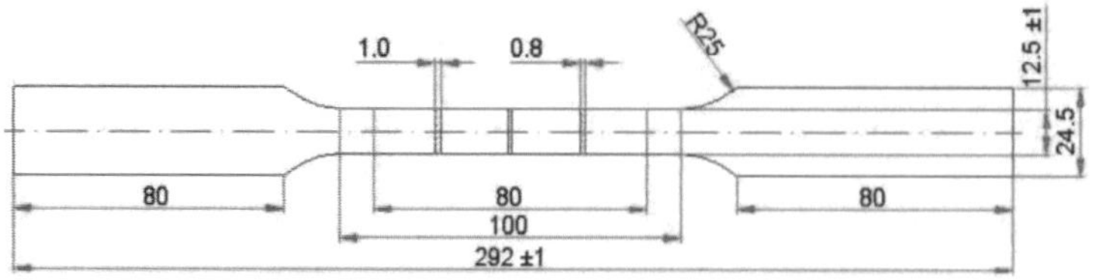

Figure 1. Standard dimension specimen (A)

(a) (b)

Figure 2. Non – standard (low dimension) test specimen with (a) and without weld (b).

3. Results and discussed

3.1. *Tensile test*

Data on mechanical properties measured under tensile tests for the TWB1 and TWB2 are shown in Table 3 and Table 4. It is seen that presence of defects does not affect mechanical response substantially (see their description below). The non–standard samples („N") did not fail over the weld region. High thermal output, as well as high cooling rate, stimulates the formation of fine microstructure within the narrow width weld.

Further inspection of the weld does not reveal high porosity level. In so doing, the pores do not exert any reducing impact onto the tensile properties of the welded joint. However, it can be caused by the high hardness of the WM and HAZ regions, which effectively protects faulty (defective) weld from fast fracture.

Table 3. Average values of mechanical properties (Base metal/TWB1).

	Base metal DX54D+Z – 0.8 mm			Welded joint - TWB1		
Mechanical properties	v = 30 mm.min^{-1}A	v = 0.9 mm.min^{-1}B	v = 30 mm.min^{-1}B	v = 30 mm.min^{-1}A	v = 0.9 mm.min^{-1}B	v = 30 mm.min^{-1}B
σ_{ys} [MPa]	192 ^{S}	173	200	187 ^{S} 183* -184	189* 192	217
σ_{ts} [MPa]	291 ^{S}	275	295	298 ^{S} 291* -297	287* 289	304
ε_t [%]	37.7 ^{S}	31.9	30.9	27.2 ^{S} 21.2* 27.2	21.2* 26.2	22.3

Table 4. Average values of mechanical properties (Base metal/TWB2).

	Base metal DX53D+Z - 1.3 mm			Welded join – TWB2		
Mechanical properties	v = 30 mm.min^{-1}A	v = 0.9 mm.min^{-1}B	v = 30 mm.min^{-1}B	v = 30 mm.min^{-1}A	v = 0.9 mm.min^{-1}B	v = 30 mm.min^{-1}B
σ_{ys} [MPa]	312 ^{S}	311	338	321 ^{S} -319	336.67	347
σ_{ts} [MPa]	389.6 ^{S}	376.7	400	396 ^{S} -392	385	396
ε_t [%]	24.9 ^{S}	20.3	24.2	19.1 ^{S} -19.03	15.33	17.9

S – standard dimension test specimens (A); (-) – standard dimension test specimens with damaged welds

Radiometric inspection possesses high sensitivity to reveal faults in weld metal. However, the tensile tests performed on the non-standard and standard size sample showed low sensitivity to presence of faults. In so doing, high strength of the weld supresses development of faults. The variation of key mechanical properties was evaluated through measurement of Tensile strength/Yield strength and elongation at break. The results obtained evidence that with increasing loading rate, a slight increase in the Tensile strength/Yield strength and elongation at break takes place.

3.2. *Microhardness and metalograph*

Average values of microhardness measured for the TWB1 tensile test specimens no. 3 and no. 4 are shown in Table 5. Average values of microhardness measured for the TWB2 tensile test specimens no. 1 and no. 2 are shown in Table 6. The data of their optical microscopy observation is illustrated by figure 3 and figure 4. Microstructure of the base metal DX54D+Z represents ferrite pattern and grains of the polyedrical shape. The DX53D+Z has ferrite-cementite structure; however, cementite is hardly seen over the sample cross section. The laser weld metal has a characteristic dendrite structure.

Table 5. Average values of microhardness of TWB1.

No.	Location of measurement	Specimen no.3	Specimen no.4
1	Base metal 1 DX53D+Z (1 mm)	106	110
2	HAZ1 DX53D+Z (1 mm)	122	125
3	WELD	229	256
4	HAZ 2 DX54D+Z (0.8 mm)	111	118
5	Base metal 2 DX54D+Z (0.8 mm)	102	104

Table 6. Average values of microhardness of TWB2.

No.	Location of measurement	Specimen no.1	Specimen no.2
1	Base metal 1 DX53D+Z (1.75 mm)	112	108
2	HAZ1 DX53D+Z (1.75 mm)	126	129
3	WELD	235	242
4	HAZ 2 DX54D+Z (1.3 mm)	125	124
5	Base metal 2 DX54D+Z (1.3 mm)	108	113

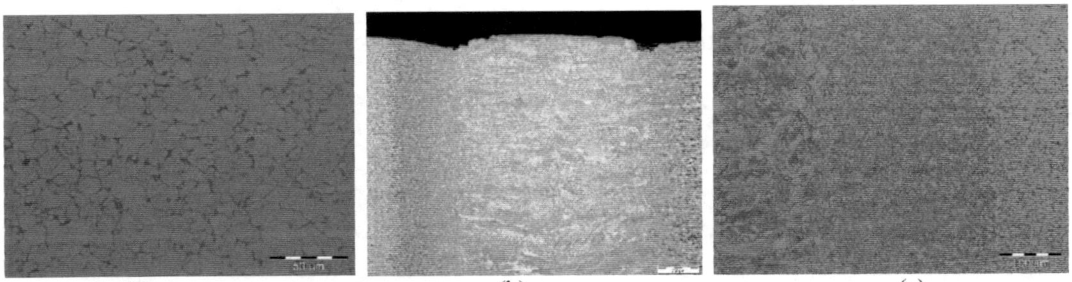

(a) (b) (c)

Figure 3. Structure of welded joint TWB1: a - base metal DX53D+Z; b – weld; c – HAZ 2 DX54D+Z.

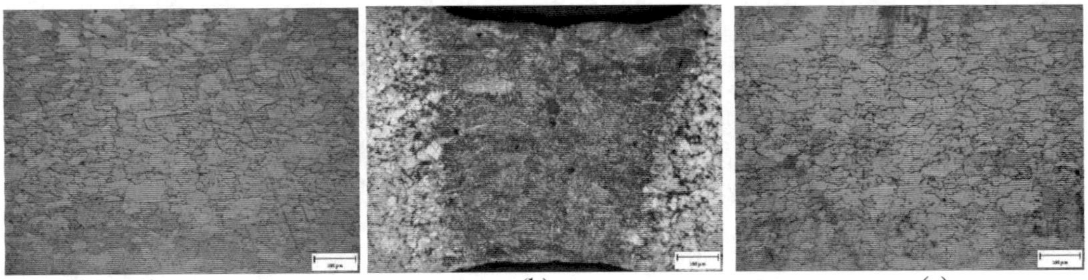

(a) (b) (c)

Figure 4. Structure of welded joint TWB2: a - base material DX53D+Z; b - weld and HAZ; c - base material DX54D+Z.

The structure of the weld has a gradient pattern. It consists of a fusion zone; an overheated zone is located on the boundary of the zone of molten metal, which was not subjected to melting, and a normalization zone with a fine-grained structure. The fusion zone is formed by round-shaped grains, figure 4a, oriented at an angle to the center of the weld. In general, microalloyed steels are less

IOP Conf. Series: Materials Science and Engineering **731** (2020) 012005 doi:10.1088/1757-899X/731/1/012005

incoherent over the austenite phase growth in the HAZ because of its blocking by the precipitates. Most dangerous effect of structure modification consists in increasing the size of the overheating zone, especially in the upper part of the weld, since the welded metal has the lowest strength properties there. The thin sheets, i.e. TWB1, tends to be more sensitive to the welding parameters.

Part of the study was dedicated to monitoring the uprising defects, i.e. pores in the weld. Figure 5 shows the pore in the region where the EDX analysis was conducted. However, diffusion of Zn from the protective layer into the weld metal is not revealed. An increased content of oxygen and aluminium was detected at the pores [1]. In the TBW1 sample, some particles become evident in the weld. The reason is related to the edge treatment procedure (grooving). Note that edges were not properly treated before the welding [16]. Welding of the thin sheets (TWB1) exhibits high sensitivity to the width of the weld. In doing so, air coming from the voids was the main reason for the nucleation of plenty of pores.

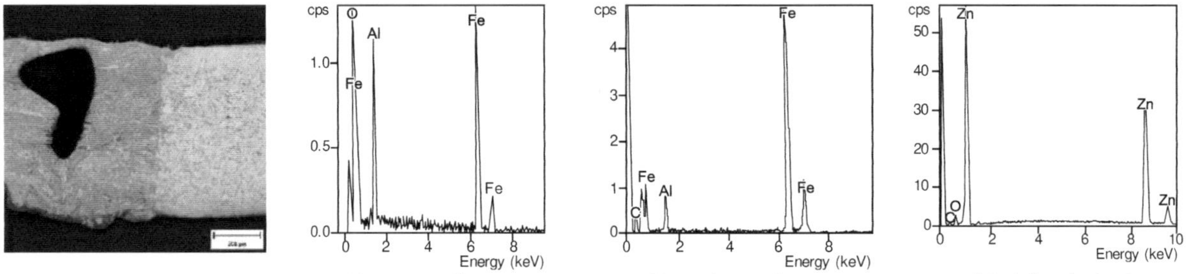

Bottom of a pore | Pore boundary | Metal substrate

Figure 5. Micrograph of a pore and data of the EDX analyses.

When ensuring higher welding quality for the TWB2, the nucleation of pores was revealed. However the level of defectiveness was admissible (ranking to 1 – 3 points). At the same time, the 80% of TWB1 welds contains too many defects and their level of defectiveness is estimated as non-acceptable.

With the use of penetrant inspection, pores were revealed on the surface of welds. For steels (DX53D+Z and DX54D+Z), they mainly comes from aluminum oxides and nitrides [17]. Gas porosity and gas bubbles can also arise due to a release of gases dissolved in the metal during welding process.

A decrease in the welding speed while maintaining other parameters of the process eliminates shrinkage cavities primarily in the upper part of the weld. However, it gives rise to nucleation of an equally dangerous defect - the formation of a penetration zone with an unstable width of the weld. At the same time, an active formation of defects in the heat affected zone is normally accompanied by the redistribution of chemical elements and even dispersed phases, which in turn stimulates the processes of phase formations in the metal [18, 19]. Moreover, phase formations of certain sizes can "strengthen" grain boundaries, contributing to the refinement of the structure.

Laser welding is one of the promising methods of car bodies manufacturing. But since the nucleation of defects such as pores, holes, lack of penetration cannot be excluded completely, the post-processing inspection of products should be obligatory applied.

4. Conclusion

The key factors of the technological parameters of laser welding of zinc-coated steel sheets (DX53D + Z and DX54D + Z) plates affecting structure formation as well as strength of the weld have been determined. It is found that the weld structure has a gradient pattern. Technological recommendations are formulated to reduce the defectiveness of the structure in the primary crystallization zone as well as the overheating zone, based on the morphology analysis of the resulting pores and weld as a whole.

Acknowledgement
This work was supported by scientific grant agency of the Ministry of Education of the Slovak Republic VEGA No. 1/0424/17, KEGA 001STU-4/2019 and of the Slovak Research and Development Agency APVV-16-0359. S. Panin acknowledges financial support of the program of basic scientific research of state academies of sciences for 2013-2020, line of research III.23.

References
[1] Viňáš J 2000 Laser technologies used in welding of surface treated materials *Diploma thesis, Košice*
[2] Razmpoosh M H, et al. 2018 *Materials & Design* **155** pp 375-383
[3] Deng S et al. 2019 *Int. J. of Heat and Mass Transfer* **140** pp 269-280
[4] Sejč P and Világoš T Strojnicky Casopis 2018 68 pp 89-94
[5] Gaoyang M, Lingda X, Chunming W, Ping J and Guoli Z 2019 *Materials & Design* **181** Art. number 107980
[6] Čičo P, Kalincová D and Kotus M 2011 *Research in Agricult. Eng.* **57** pp 50-55
[7] Wahba M, Kawahito Y and Katayama S 2011 *30-th Int. Cong. on Applications of Lasers and Electro-Optics* ICALEO pp 343-349
[8] Zdravecká E and Slota J 2019 *Metals* **9** (91) https://doi.org/10.3390/met9010091
[9] Kim J, Oh S, Son M and Ki H 2017 *J. of Mat. Proc. Tech.* **249** pp 135-148
[10] Hong K-M, Shin YC, 2017 *J. of Mat. Proc. Tech.* **245** pp 46-69
[11] Kim J, Oh S and Ki H 2016 *J. of Mat. Proc. Tech.* **232** pp 131-141
[12] Fengde L et al. 2019 *Mat. Research Express* **6** Art. number 076505
[13] Vyskoč M, Sahul M and Sahul M 2018 *J. of Mat. Eng. and Performance* **27** pp 2993-3006
[14] Sahul M, Turna M and Sahul M 2014 *Magnesium Technology 2014 - TMS 2014 143rd Annual Meeting and Exhibition* San Diego CA United States pp 301-305
[15] Viňáš J and Ábel M 2015 *Materials Science Forum, International Conference on Surface Engineering and Materials in Mechanical Engineering* High Tatras Slovakia **818** pp 239-242
[16] Vuherer T, Maruschak P O and Samardžić I 2012 *Metalurgija* **51** pp 301-304
[17] Viňáš J, Brezinová J, Brezina J and Maruschak P O 2019 *Materials Science* **55** 1 pp 46-51
[18] Konovalov S, Kormyshev V, Gromov V and Ivanov Yu 2016 *Mater. Sci. Forum* **870** pp 159-162

The Effect of Plastic Deformation on the Geometric Parameters of Nb-Ti-Based Superconducting Cable

A Yntymakova[1,2] and G Shlyakhova[2]

[1]National Research Tomsk Polytechnic University, Tomsk, Russia
[2]Institute of Strength Physics and Materials Science SB RAS, Tomsk, Russia

E-mail: aklima.int@mail.ru

Abstract. The current carrying capacity of a superconductor depends on its microstructure, that is, the shape, size and volume content of particles released during the decomposition of a solid solution. At the technological stage, it is necessary to ensure continuity of the process and a given density of micro defects. Therefore, the aim of this work was to assess and present the results of studies of the effect of plastic deformation (tension) on the structure of a multicore superconductor based on Nb-Ti alloy. Mechanical tests were carried out for uniaxial tension to failure on a Walter + Bai AG LFM-125 testing machine (maximum force up to 125 kN). The internal structure of the sample at the fracture site after the sample preparation was studied using a Neophot-21 optical microscope. The effect of deformation on the geometric parameters of Nb-Ti fibers was studied. Sample preparation included thin sections, polishing and etching.

1. Introduction

Superconductivity is one of the most outstanding discoveries of the twentieth century and lies in the ability of certain substances to conduct electricity without resistance at very low temperatures, that is, become superconductors [1-2]. The critical temperature of the superconductor based on Nb-Ti alloy is 9.8 K, the critical field is 12 T. The most important applications of superconductors are the creation of strong magnetic fields, generation and transmission of electricity. A solenoid made of a superconducting material can work without supplying energy from outside, until at some point in its life the solenoid valve fails and stops working.

Maintaining a solenoid in a superconducting state does not require large energy costs. With zero resistance, the problem of heat dissipation is easily solved. In addition, superconducting magnets are much more compact than usual. Each kilogram of the mass of the superconducting magnet creates a magnetic field equivalent in strength to the field of a 20-ton electromagnet with an iron core.

The problems of thermonuclear energy cannot be solved without the use of powerful superconducting magnets. To carry out controlled thermonuclear fusion of helium nuclei from deuterium and tritium nuclei, it is necessary to maintain hot tritium-deuterium plasma heated to 10^8–10^9 °C in the reaction space. Only superconducting magnets can create fields of this force. Superconductors are used to produce magnetic fields with large induction. For example, superconducting alloys with large critical inductance are used to make wire for transformer windings. High-density current is generated in such windings and the electromagnet has a magnetic field of great strength. The induction of the resulting fields reaches 10 T. Alloys and compounds of niobium become superconducting at sufficiently high temperatures. They can withstand fairly strong magnetic fields and are characterized by a high current

IOP Conf. Series: Materials Science and Engineering **731** (2020) 012006 doi:10.1088/1757-899X/731/1/012006

density. In liquid helium with an external field with an induction of 2.5 T, the critical current density for Nb-Ti is 2.5 kA / mm^2 [3-4].

2. Materials and methods of research

Mechanical tests for uniaxial tension to failure were carried out on two identical samples ø 1.3 mm on a Walter + Bai AG LFM-125 testing machine (maximum force up to 125 kN). An example of a sample in its initial state is shown in Figure 1, a cross sectional image was obtained using Neophot-21 optical microscope with 100x magnification [5]. As shown in the figure: the cable consists of a Cu sheath, a Cu matrix and about 3,500 Nb-Ti fibers located between them and each fiber is in the niobium barrier. Niobium barriers serve to prevent the diffusion of the Nb-Ti alloy into the stabilizing copper sheath and the occurrence of stray currents. All Nb-Ti strands have round-shaped cross section and the following chemical composition: 35.66 at. % Nb and 63.07 at. % [6-10].

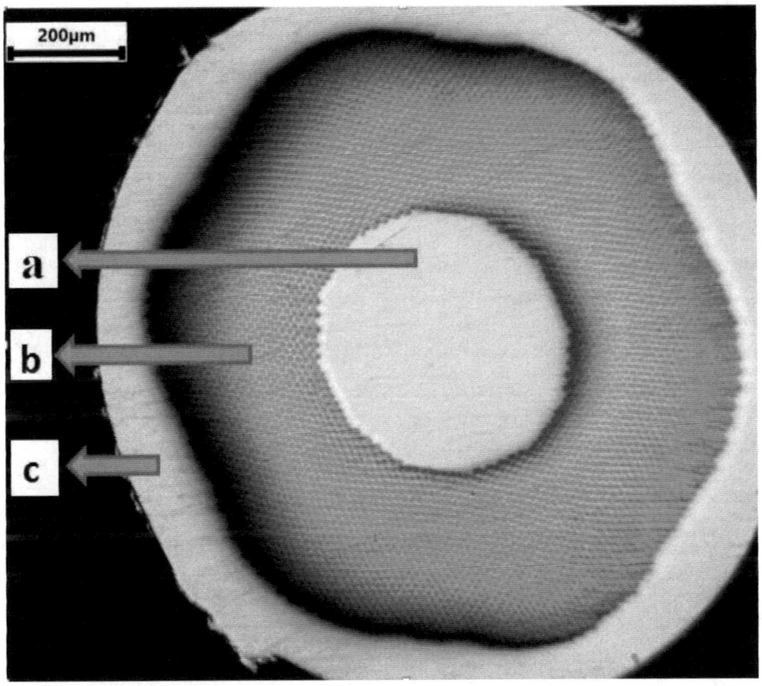

Figure 1. Cross sectional image obtained in the initial state: (a) copper matrix, (b) Nb-Ti fibers, (c) copper sheath.

Figures 2 and 3 show diagrams of tensile specimens during mechanical tests for uniaxial tension to failure on a testing machine. Figure 2 shows a diagram of a sample with a strain of ε = 3.4 % the elongation at break was 1.79 %; the fracture necking was 2 %. For the sample with strain ε = 4.3 % the elongation at break was 2.21 %; the fracture necking was 2.50 %.

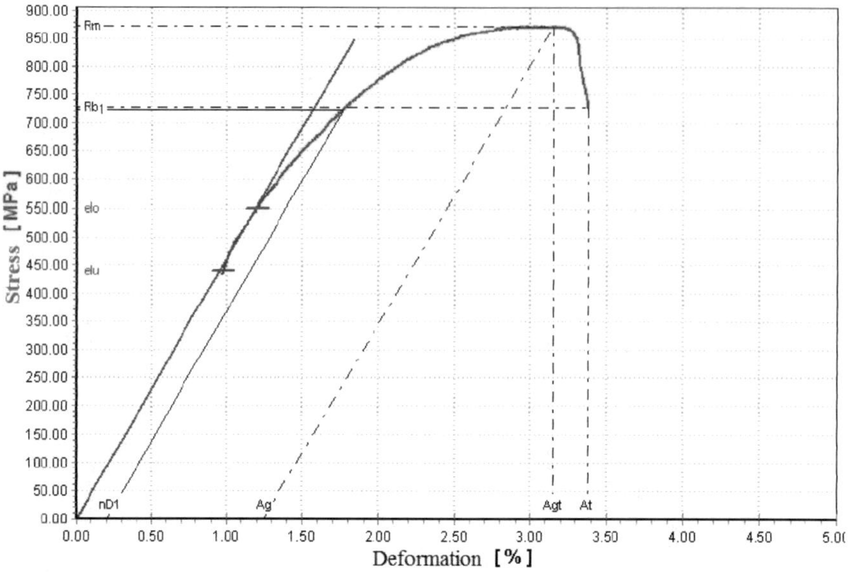

Figure 2. Fracture deformation: 3.4%; the green line indicates the elastic limit, the red line indicates the irreversible deformation.

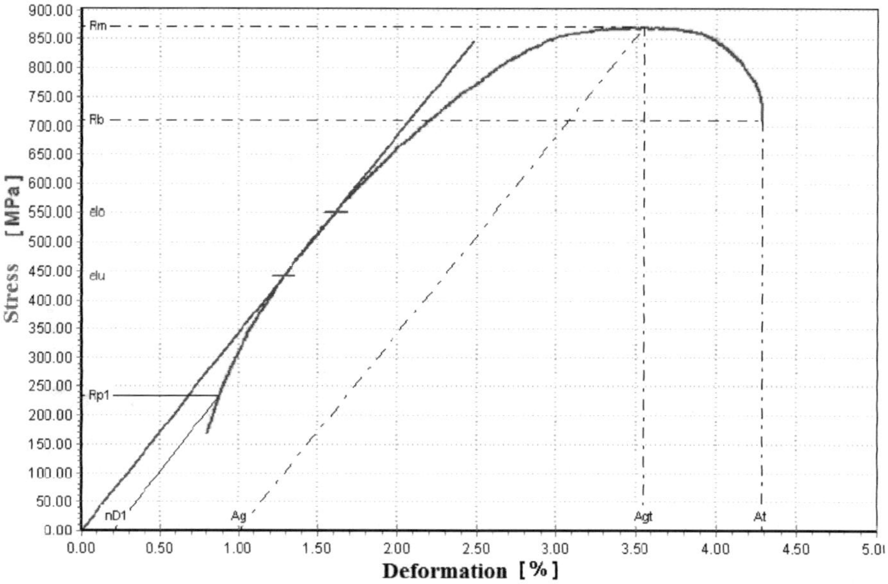

Figure 3. Fracture deformation: 4.3%; the green line indicates the elastic limit, and the red line indicates the irreversible deformation.

3. Results

To study the structure of the superconductor after tensile tests, metallographic thin sections were prepared according to standard procedures for these alloys [11]. Finally, thin sections were etched with reagent [12].

Figures 4 and 5 show images of the structures of the samples after breaking at various degrees of deformation $\varepsilon = 3.4$ % and $\varepsilon = 4.3$ %. The influence of deformation on the geometric parameters was studied and the increase in the deformed fibers from the copper core to the shell was calculated: the region of deformed fibers in the sample with deformation ε = 3.4 % was 35 %, in the sample with deformation ε = 4.3 % it was 58.5 %. Images were obtained using an optical microscope Neophopt-21 with 500x magnification.

Figure 4. Sample with ε = 3.4%. The region of the located fibers from the matrix to the cable sheath is 335 μm.

As can be seen in the panoramic figures, in the region of the copper shell of the sample, Nb-Ti fibers had an elongated square or rectangular shape. This deformation of the fibers is explained by the load on the fibers under tension precisely on the edges of the sample. Moreover, in the region of the copper matrix, the fibers had a less deformed round shape. In the initial state, the regular shapes of the fibers are round.

Figure 5. Sample with ε = 4.3%. The length of the fibers located from the matrix to the cable sheath is 350 μm.

4. Conclusion

After studying the sample it was found that for the same tensile strength (R_m = 873 MPa) for samples with ε = 3.4 % and ε = 4.3 %, the elongation (δ) upon failure turned out to be different and amounted to ≈1.8 % and 2 %, respectively.

The studies showed that for the sample with deformation ε = 3.4 % the region of deformed fibers from the matrix to the cable sheath was 35 %, while for the sample with deformation ε = 4.3 % the region of deformed fibers from the matrix to the cable sheath was 58.5 %. These results evaluate the effect of plastic deformation on the geometric parameters of the sample.

IOP Conf. Series: Materials Science and Engineering **731** (2020) 012006 doi:10.1088/1757-899X/731/1/012006

Future aspect of this investigation is very useful for industrial applications and can be used in further research work.

Acknowledgments
This work was performed as part of the Fundamental Research Program of the State Academy of Sciences for 2013–2020, line of research III.23.

References
[1] Bardeen J, Cooper L N and Schrieffer J R 1957 *Theory of superconductivity (*American Physical Society) p 1175–1204
[2] Kalimov A G 2007 Physical foundations of *superconductivity J. St. Petersburg: RADIO AND COMMUNICATIONS* 344-52
[3] Shikov A V 2003 The development of superconductors for the ITER magnetic system in Russia *J. Non-ferrous metallurgy* 1
[4] Soloviev V 2017 International Thermonuclear Reactor (ITER) *J. Spacegid (Electronic Materials)*
[5] Vlasov A I, Elsukov K A and Kosolapov I A 2011 *Optical Microscopy* A Training Manual. Publishing House of MSTU named after Bauman N E p 184
[6] Lee P J, Larbalestier D C and McKinnel J C High titanium 1988 Nb-Ti alloys initial microstructural studies *Adv. Cryog. Eng. Mat. 34* p 967-74
[7] Cherny O V, Andrievskaya N F and Ilicheva V O 2002 The microstructure and critical current density of Nb-48 wt.%Ti superconductor with very high alpha-Ti precipitate volume and very high critical current *Adv. Cryog. Eng. 48B (2002)* p 883-90
[8] Cherny O V, Storozhilov G E and Andrievskaya N F 2005 Structure and properties of differently directed deformed niobium-titanium alloy *IEEE Trans. Appl. Supercond.* 15 p 3502-05
[9] Zuev L B, Shlyakhova G V, Barannikova S A and Kolosov S V 2013 Microstructure of elements of a superconducting alloy Nb-Ti cable *Russ. Metal* p 229-34
[10] Zuev L B, Shlyakhova G V, Barannikova S A and Malinovskiy A May 2016 Microstructure of superconducting cable components *Int. J. GEOMATE* Vol 10 p 1906-11
[11] Karpov Y A 2015 Methods of sampling and sample preparation *(Electronic Materials, manual 3rd ed.)* Laboratory of Knowledge p 246
[12] Becker M and Klemm H 1988 Methods of metallographic etching Handbook *M. Metallurgy*

IOP Conf. Series: Materials Science and Engineering **731** (2020) 012007 doi:10.1088/1757-899X/731/1/012007

Comparative Analysis of the Polymeric Materials Deformation Behavior under Squeezed and Free Compression

A A Adamov[1], A A Kamenskikh[2] and Yu O Nosov[2]

[1]Institute of Continuous Media Mechanics, Perm, Russia
[2]Perm National Research Polytechnic University, Perm, Russia

adamov.aa@ya.ru, anna_kamenskih@mail.ru, ura.4132@yandex.ru

Abstract. The results of experimental studies of the modern antifriction polymer materials and composites based on them are presented in the work. The 6 most promising materials for use as antifriction coatings and interlayers in friction units of critical structures were selected from a set of materials: bearings, expansion joints etc. The deformation theory of elastic-plasticity for the active loading case is chosen to describe the polymer behavior model based on the experiments results in a first approximation. Compression charts are defined for the selected materials. Numerical models of constrained and free compression experiments are constructed. The series of numerical experiments established that the contact pressure level of composite materials is 25-40% lower at a single level of deformation. UHMWPE and modified PTFE under uniaxial deformed conditions state have an almost elastic behavior with weakly nonlinear hardening and in test problems their deformation behavior has small differences.

1. Introduction

A large set of polymer materials of Russian and foreign production suitable for use as antifriction coatings and interlayers in structural units operating in the framework of contact interaction with friction. Such materials are widely used in aviation technology [1], construction [2], medicine [3], engineering [4-5] and other fields. The modern antifrictional polymer materials include: antifriction composite polymer materials based on PTFE [6-9], modified PTFE [8-10], ultra-high molecular weight polyethylenes (UHMWPE) of Russian and foreign production and composite materials based on them [6, 8, 11]. Information on the materials properties and structure is necessary to solve the deformation behavior problems of structures with antifriction coatings and interlayers [6, 12]. Russian and foreign scientists are studying the physicomechanical, chemical, frictional and rheological properties of modern polymeric materials and composites based on them [6-18 and etc.]. Information insufficiency about the materials properties under study is still noted [16, 18], which hinders their effective application in many industries. Experimental study of the polymers properties used as antifrictional layers and the numerical models construction of their behavior in the framework of contact mechanics are relevant areas of research. Interdisciplinary studies of the modern polymers behavior are made in the work and include a series of experiments, processing of experimental data with the modern polymer behavior models construction and numerical simulation of the studied materials deformation behavior on test problems.

2. Experiments

The Institute of Continuous Media Mechanics of the Ural Branch of the Russian Academy of Sciences carried out a cycle of experimental studies of the physical and mechanical characteristics of antifriction materials at complex multistage deformation stories with offloading using a Zwick Z100SN5A testing machine. A series of full-scale experiments included (figure 1): tests to determine the Brinell hardness of materials by penetrating a ball with a diameter of 5 mm; studies under free compression conditions, as well as studies of constrained compression by pressing cylindrical specimens with a diameter and length of 20 mm in a special fixture with a rigid steel cage.

Figure 1. Experimental studies of antifriction polymers: a) Brinell hardness, b) uniaxial stress state; c) uniaxial deformed conditions state.

More than 30 modern antifrictional materials and composite materials based on them, suitable to one degree or another as sliding interlayers in friction units were investigated in the experiments series. The 6 modern antifrictional polymers and composites based on them were selected from a wide range of materials as the most promising for applicaation in friction units: carbon filled UHMWPE (Material 1); UHMWPE of Russian production (Material 2); UHMWPE of Germany production (Material 3); antifriction composite material based on PTFE with a dendrite bronze inclusions and molybdenum disulfide (Material 4); antifriction composite material based on PTFE with a spherical bronze inclusions and molybdenum disulfide (Material 5); modified PTFE (Material 6).

The tangent modulus M and elastic modulus E were determined from the test results at free and constrained compression. Other elastic constants of an isotropic elastic body are expressed through measured modules according to [19]. Poisson's ratio is defined by the formula $\nu = E/2\mu - 1$ where $\mu = E/8\left(1 + 3M/E - \sqrt{(1+3M/E)^2 - 16M/E}\right)$. The elastic mechanical characteristics for the selected set of modern antifriction materials are presented in the table 1.

Table 1. Properties of modern antifrictional materials

№	Antifrictional materials	E (MPa)	ν	№	Antifrictional materials	E (MPa)	ν
1	Material 1	1420.00	0.4402	4	Material 4	903.00	0.4465
2	Material 2	1050.00	0.4699	5	Material 5	860.52	0.4388
3	Material 3	706.00	0.4522	6	Material 6	863.80	0.4610

It was found that modern antifrictional polymeric materials and composites based on them exhibit nonlinear properties. The deformational theory of elastic-plasticity is chosen to describe the behavior model of the materials as a first approximation (figure 2).

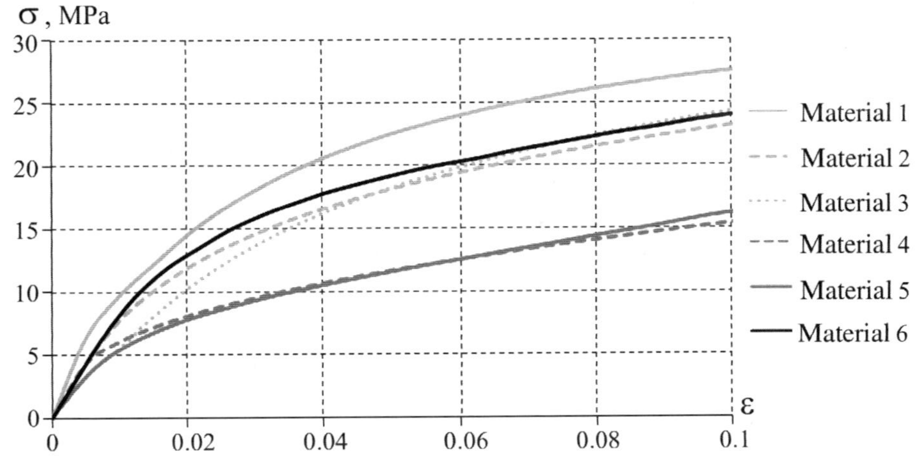

Figure 2. Compression diagrams $\sigma-\varepsilon$.

Compression diagrams $\sigma-\varepsilon$ obtained experimentally at low strain rates or determined by constructing envelope curves when processing cyclic free compression diagrams are shown in figure 2. Numerical calculation using the selected material behavior model gives good quantitative agreement with the experimental results for the case of active loading what is established in [17].

3. Mathematical modeling

The numerical simulation of experiments on constrained and free compression of cylindrical samples was performed as part of the influence analysis of the modern antifriction materials properties. The boundary-value deformation elastoplasticity problem in an axisymmetric formulation taking into account frictional contact interaction over the mating surfaces of the press plates and cylindrical samples is realized (figure 3). Contact interaction with a rigid steel cage is not taken into account. Squeezed compression of samples is shown in the scheme 1 and free compression of samples is shown in the scheme 2.

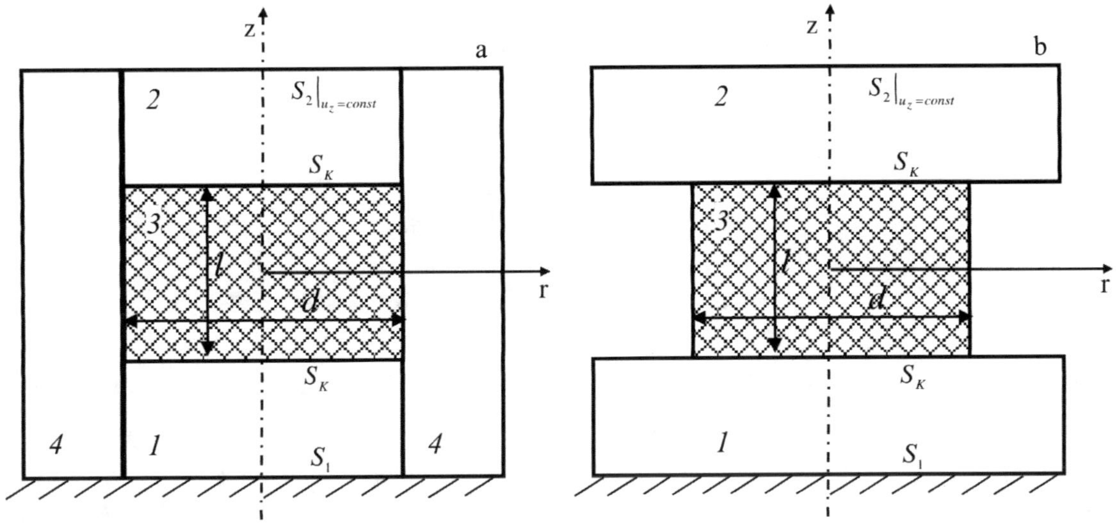

Figure 3. Numerical model of the experiment: a) constrained compression; b) free compression.

The contact problem mathematical formulation of an elastoplastic polymeric material with press plates was previously described in [17-18] and is supplemented by the following boundary conditions:

– on the surface S_1 $\vec{u}=0$, $\vec{x}\in S_1$;

– on the surface S_2 $u_z=\Delta l$, $\sigma_{r\varphi}=0$, $\vec{x}\in S_2$;

– the other external surfaces are free $\hat{\sigma}\cdot\vec{n}=0$.

The polymer-metal friction coefficient is selected from the reference literature and is 0.04.

4. Analysis of the study results

An analysis of the numerical solution problem convergence from the system discretization degree at a cylindrical sample maximum deformation of 10% as an example of material 6 was performed as part of the work. Seven variants of finite element mesh with a decreasing gradient of finite element size to the contact area is considered. A finite element mesh with a maximum element size $l/200$ on the material main volume and the minimum element size $l/600$ near the contact area is selected as optimal. This finite element partitioning provides optimal accuracy and computation time solution.

The influence of the antifriction polymeric materials properties on the contact parameters with an increase in the sample deformation to 10% was considered on the finite element mesh of selected version. It was found that contact pressure and contact tangential stress do not change significantly for the squeezed compression case on the contact surface of main volume. The maximum peaks of contact parameters are observed near the edge of cylindrical sample. This effect is associated with the change of the contact status with the "adhesion" on "sliding". The level of the contact pressure and contact tangential stress rises to the sample edge smoothly by approximately on 25-30% in the case of free compression. Peaks of contact parameters are observed near the sample edge similarly to the first calculation scheme.

The average values of contact zone parameters at the maximum strain level of 10% for two design schemes versions are presented in figure 4.

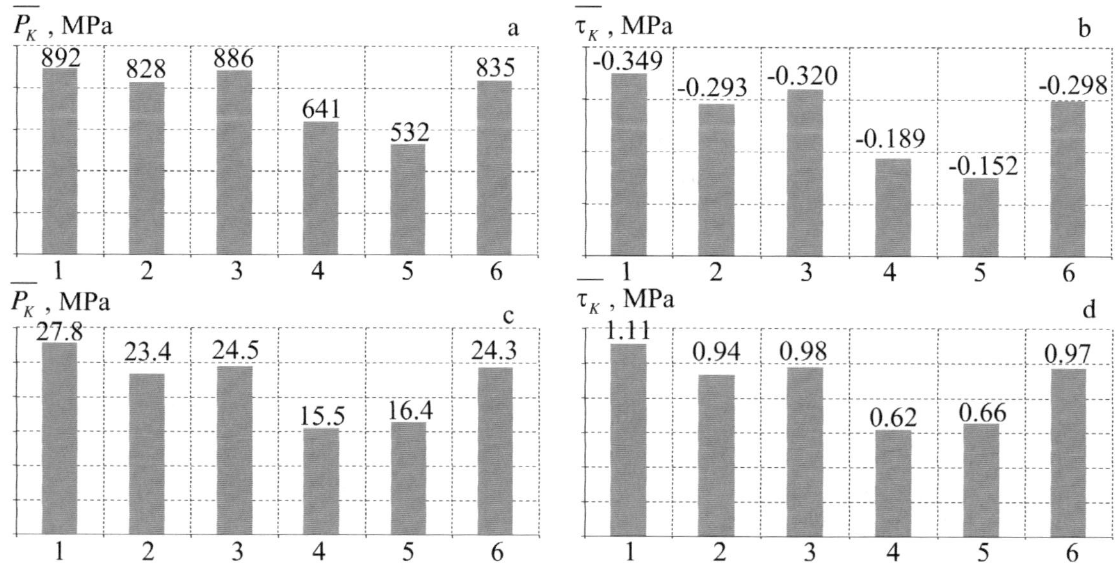

Figure 4. Average values of contact press (a, c) and tangential stress (b, d) on S_{K_1} at $\varepsilon=10$ %: a, b for scheme 1; c, d for scheme 2; 1-6 are materials 1-6 respectively.

All sample of the UHMWPE and modified PTFE at 10% strain have small differences between the contact pressure level: within 8 and 20% for the case of squeezed and free compression respectively. Deformation of samples (material 4-5) equal to 10% is achieved at significantly lower levels of contact pressure and contact tangential stress than other materials considered: in the case of constrained compression contact pressure level is lower by 25-40% and the contact tangential stress is below by 20-35%; for the case of free compression the level of contact pressure and tangential stress is lower by 35-40%.

The dependences of the maximum level of contact parameters modulo on the cylindrical samples deformation under squeezed compression are shown in figure 5.

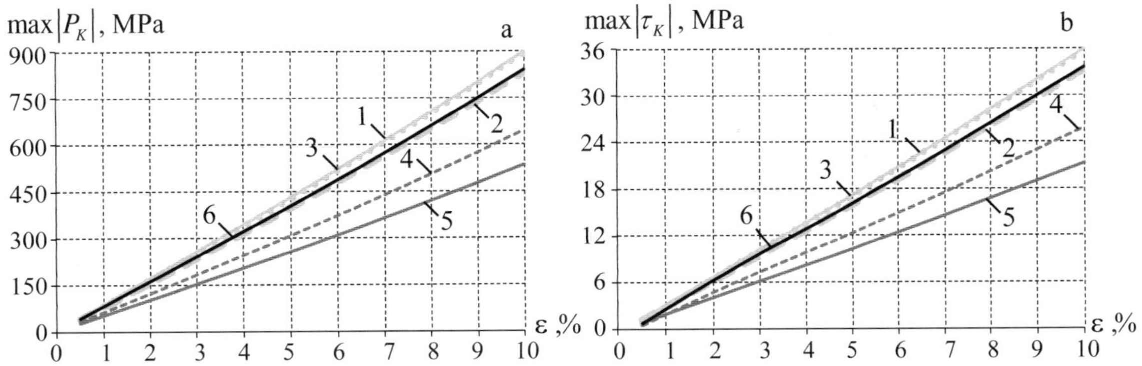

Figure 5. The dependences of the maximum modulo level of contact press (a) and tangential stress (b) on surface S_{K_1} on ε (design scheme 1): 1-6 are materials 1-6 respectively.

The dependence of the maximum level of contact pressure and contact tangential stress on the samples deformation is close to a linear law. At one level of contact parameters the deformation of samples from modern antifriction composite materials is 25-30% more than that of the other considered polymers. And $\max|P_K|$ is 0.5-0.7% higher than the average contact pressure $\overline{P_K}$. The maximum level of contact shear stress is on average ~ 25 times less than contact pressure.

The dependences of the maximum level of contact parameters modulo on the cylindrical samples deformation under free compression are shown in figure 6.

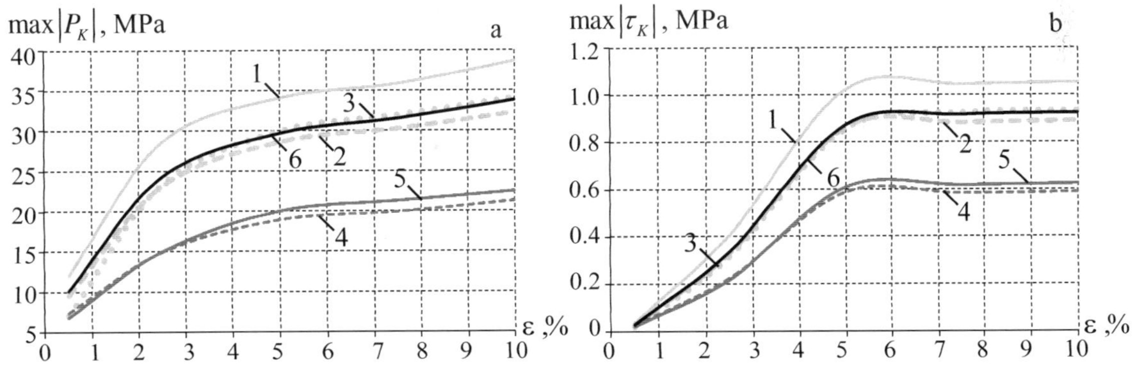

Figure 6. The dependences of the maximum modulo level of contact press (a) and tangential stress (b) on S_{K_1} at ε (design scheme 2): 1-6 are materials 1-6 respectively.

The dependence of the maximum level of contact pressure and contact tangential stress on the samples deformation is not linear. The maximum contact pressure level of design scheme 2 is approximately 95-96% lower than for design scheme 1. At one level of contact parameters the deformation of samples from modern antifriction composite materials is 30-45% higher than that of the other considered polymers similar to the first design scheme. The maximum level of contact tangential stress is on average ~ 35 times less than contact pressure.

5. Conclusion

A series of experimental studies of polymer antifriction materials aimed at obtaining data on their physical and mechanical properties was performed as part of the work. The studies were performed for a wide range of modern antifriction polymer materials and composites based on them. The elastic-plasticity deformation theory is chosen to describe the behavior model of materials in a first approximation, Young's modulus and Poisson's ratio of the elastic portion are determined, diagrams $\sigma-\varepsilon$ are described.

As part of the deformation behavior analysis of modern antifriction materials it has been established:

- carbon filled UHMWPE has the best deformation characteristics, but additional experimental studies with various histories of prolonged multi-stage loading are needed to build and identify a qualitative model of its behavior;
- the other UHMWPE's under consideration and the modified PTFE in the uniaxial deformed state exhibit almost elastic behavior with weakly nonlinear hardening, in test problems their deformation behavior has small differences;
- antifriction composite materials have more significant non-linear and viscoelastic effects, as well as greater pliability compared to other materials studied, rheological effects are more noticeable with dendritic particle shape.

The numerical modeling results do not contradict to the experiments results.

Acknowledgments
This work was supported by the RFBR (grant № 18-08-00903).

References

[1] Rakowski W A and Zimowski S 2006 *Compos.: Part B Eng.* **37** 81-88
[2] Pinchuk L S, Nikolaev V I, Tsvetkova E A and Goldade V A 2006 *Tribology and biophysics of artificial joints* (Amsterdam: Elsevier) p 350
[3] Tukashev Zh B and Adilhanova L A 2013 *Geol., Geography and Global Energy* **2** 163-166
[4] Wu Yi, Wang H, Li Ai, Feng D, Sha B and Zhang Yu 2017 *J. of Zhejiang University-Sci. A.* **18(5)** 363-376
[5] Peel H, Luo S, Cohn A G and Fuentes R 2018 *Autom. in Construction* **94** 244-256
[6] Yi X-S, Du S and Zhang L 2018 *Composite Materials Engineering, Volume 1: Fundamentals of Composite Materials* (Singapore: Springer) p 765
[7] Solonin Yu M, Nenakhov A V, Kostornov A G, Danilenko N I, Gorban' V F and Karpets M V 2014 *Powder Metal. and Metal Ceram.* **52(11-12)** 620-631
[8] Kamenskih A A and Trufanov N A 2013 *Comput. Continuum Mech.* **6(1)** 54-61
[9] Shelestova V A, Grakovich P N, Danchenko S G and Smirnov V A 2006 *Chem. and Pet. Eng.* **42(11-12)** 663-666
[10] Martynov M A, Gol'dman A Ya, Il'chenko P A, Polyakov Yu S and Andreeva A I 1975 *Strength of Mater.* **7(11)** 1390-1393
[11] Pavlenkoa V I, Bondarenkob G G, Tarasova D G and Edamenko O D 2013 *Inorg. Mater.: Appl. Res.* **4(5)** 389-393
[12] Ushakov A V, Karpov I V, Fedorov L U and Lepeshev A A 2014 *J. of Friction and Wear* **35(1)** 7-11
[13] Ermakov S F and Myshkin N K 2018 *Liquid-crystal nanomaterials: tribology and applications* (Cham: Springer) p 227
[14] Balyakina V B, Khatipov S A and Pilla C K 2015 *J. of Friction and Wear* **36(4)** 346-349.
[15] Ginzburg B M, Tochil'nikov D G, Shepelevskii A A, Leksovskii A M and Tuichiev Sh 2006 *J. of Appl. Chem.***79(9)** 1518-1521
[16] Adamov A A 2013 *Compos. Mater. Constructions* **2** 28-37
[17] Kamenskih A A and Trufanov N A 2015 *J. of Friction and Wear* **36(2)** 170-176
[18] Adamov A A Kamenskikh A A and Nosov Yu O 2019 *Int. J. of Civ. Eng. and Technol.* **10(5)** 99-107
[19] Adamov A A, Matveenko V P, Trufanov N A and Shardakov I N *Applied viscoelastic methods* (Ekaterinburg: IMSS UrO RAN) p 411

IOP Conf. Series: Materials Science and Engineering **731** (2020) 012008 doi:10.1088/1757-899X/731/1/012008

Research of the Materials and Reinforcement Schemes Effects on a Mechanical Behavior Smart Structure with Embedded Piezo Actuators

A N Anoshkin, P V Pisarev and V A Ashikhmin

Perm National Research Polytechnic University, Perm, Russia

E-mail: pisarev85@live.ru

Abstract. The present work is devoted to the research of a helicopter blade segment stress-strain state. The segment was made of polymer composite materials with embedded piezoelectric actuators. The computational model of the blade segment takes into account the technological order of laying and the anisotropy of the reinforcing layers properties. The reinforcement schemes under consideration were [0°/90°], [45°/-45°], [0°/45°/-45°/90°]. The static strength margin was estimated using the criterion of maximum stresses. A layered analysis of the structure stress-strain state was carried out. It was shown that the most dangerous, determining the structure margin of safety, are normal interlayer stresses. It was found that the GFRP material and the structural reinforcement scheme [0°/90°], selected during the design, provide the greatest torsion angles.

1. Introduction

Recently, in aeronautical engineering, there has been a tendency to use SMART structures with controlled geometry. This allows to significantly improve the aircraft operational characteristics by reducing the vibro-acoustic loading of highly loaded elements such as an airplane wing or helicopter blades [1]. The development of SMART structures from PCM equipped with piezoelectric elements, on the one hand, will reduce the weight of the structure, and on the other hand will eliminate mechanical drives and swashplates, which will significantly reduce the cost of servicing structures with controlled geometry and aircraft fuel consumption.

With the greatest efficiency, SMART designs can be used to create aircraft wings with variable geometry, which will optimally adapt to the aerodynamic parameters of the airflow [2]. The application of SMART technologies to damping oscillations by helicopter blades is considered. Currently, the possibility of applying SMART technologies to create fan blades and an aircraft engine straightening apparatus is being considered [3].

Currently, some works are known that describe the theoretical foundations of creating PCMSMART structures with controlled geometry [4,5]. Numerical models of the blade cross-section have been developed and parametric studies of the controlled structure equipped with AFC actuators relative stiffness have been carried out to determine the optimal configuration of the blade.

In [6], numerical results for a four-blade bearingless rotor model with self-tuning piezoelectric actuators located at an angle of 45° on the upper and lower surfaces of the composite blade were presented. To analyze the dynamic loads on the main rotor with composite blades, a complex model was developed. To model the behavior of a layered structure made of smart composite, an approach based on the theory of high-order deformation was used [7].

In [8-10], a series of works was carried out related to the analysis of the structural and aerodynamic parameters of the rotor blades in order to better understand the effect of active twisting. The influence of the blade active twisting, the required power, the load on the blade and the hub were evaluated in relation to the change in stiffness (torsional and bending in the planes of the swing and rotation), sectional mass and torsion inertia, the center of gravity and the location of the blade elastic axis. However, the scientific problem limiting the SMART constructions development (implementation in engineering design practice) is the lack of scientifically based approaches and proven methods for their design. To solve this problem, it is necessary to carry out a set of computational and experimental studies aimed at developing new mathematical models of composite materials with piezoelectric structural elements (piezocomposites), at developing methods for solving related boundary problems of electromagnetically elasticity for inhomogeneous media with piezoelectric structure elements. There is a need to design methods for calculating structures, choosing control elements, concept development, calculation methods, creating methods for designing structures, testing models and methods using structurally similar SMART structures samples, development of experimental technologies for their manufacture from modern materials, and testing these prototypes.

The present work is devoted to a controlled helicopter blade segment stress state analysis. A research was carried out of the reinforcement scheme and materials influence on the blade segment mechanical behavior using piezo thermal analogy, as well as an assessment of the strength under operational loads.

2. Numerical model

The object of this study is a controlled helicopter blade segment with embedded piezoelectric actuators. To conduct a numerical simulation of the stress-strain state, a three-dimensional computer model was developed with an explicit description of the layered structure. The geometric model includes a C-shaped spar, balancing weight, laminated Sheathing, foam filler, embedded element and piezoelectric actuators located on the blade Sheathing. At the end of the blade is a steel fastener. The NACA 23012 profile was used as the aerodynamic profile. The geometric characteristics of the helicopter blade segment are shown in Figure 1 a. The segment length is 805 mm, the chord length is 121 mm.

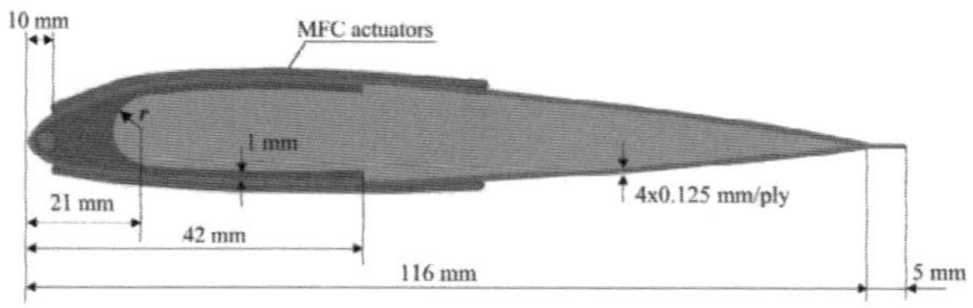

a

b

Figure 1. General view of the blade segment: a - the cross section of the blade segment; b - the geometric model of the blade segment.

The following materials were considered as blade segment materials: 1) Glass Fiber Reinforced Polymer (GFRP); 2) Carbon Fiber Reinforced Polymer (CFRP). The blade segment Sheathing consisted of four layers of GFRP or CFRP. Reinforcement schemes were [0°/90°], [45°/-45°], [0°/45°/-45°/90°].

Table 1 shows the considered variants of the blade segment samples.

Table 1. Variants of materials combinations of the blade segment samples.

№	Spar material	Sheathing material
1	CFRP	CFRP
2	CFRP	GFRP
3	GFRP	GFRP

Technical elastic constants of microfiber piezoelectric actuator (MFC) were taken from [11]:

MFC – $E_x = 30$ HPa, $E_y = E_z = 15.5$ HPa, $G_{xy} = G_{xz} = 10.7$ HPa, $G_{yz} = 5.7$ HPa, $\nu_{xy} = \nu_{xz} = 0.44$, $\nu_{yz} = 0.35$, $d_{31} = d_{32} = -1.98*10^{-10}$ m/V, $d_{33} = 4.18*10^{-10}$ m/V.

As part of the computational experiments, the mechanical behavior of helicopter blade segments equipped with piezo actuators and made of GFRP and CFRP were simulated. The maximum axial displacements of the helicopter blade segment were calculated at a control voltage level of 1000 V.

The solution of the problem was carried out by the finite element method using thermal analogy. Thermoanalogue allows to move from the need to solve the connected boundary value problem of electroelasticity to the solution of a much simpler unconnected boundary value problem of thermoelasticity. While stating the mathematical formulation, the hypothesis of the invariance of an electric inhomogeneous field in a piezoelectric element (arising from the action of a control electric voltage on its electrodes) was accepted. The hypothesis assumes the presence of a "reverse piezoelectric effect" in a piezoelectric element in the absence of a "direct piezoelectric effect" [12-14].

The strength assessment of the PCM helicopter blade segment was carried out according to the criterion of maximum stresses. This criterion was used to solve problems of this kind in [15, 16] and its choice is due to the lack of reliable data on the strength of the GFRP material under conditions of a difficult stress state.

The strength of the orthotropic material corresponds to the satisfaction of the system of inequalities at each point in the structure:

$$S_{11}^- \le \sigma_{11} \le S_{11}^+,\ S_{22}^- \le \sigma_{22} \le S_{22}^+,\ S_{33}^- \le \sigma_{33} \le S_{33}^+,\ S_{12} \le \sigma_{12},\ S_{13} \le \sigma_{13},\ S_{23} \le \sigma_{23}, \quad (1)$$

where S_{ii}^+, S_{ii}^-, S_{ij} are the limits of the material's static tensile, compression, and shear strengths, respectively, in the local coordinate system of the layer.

Safety margin of equal strength GFRP for various components of the stress state was estimated by the equation:

$$n_{ij} = \min_{\mathbf{r} \in V^{(1)}} \left(\frac{S_{ij}}{\sigma_{ij}(\mathbf{r})k} \right), \quad (2)$$

where k is the coefficient of decrease in the static strength of the material due to the influence of various environmental factors.

The n_{ij} minimum value for all σ_{ij} components will determine the margin of safety of the controlled helicopter blade segment.

3. Results of numerical modeling

Based on the results of numerical modeling of mechanical behavior under the action of a 1000 V control voltage, the fields of stresses and displacements for a helicopter blade segment made of GFRP and CFRP were obtained. Figure 2 shows the axial displacement distribution fields U_y of the blade segment.

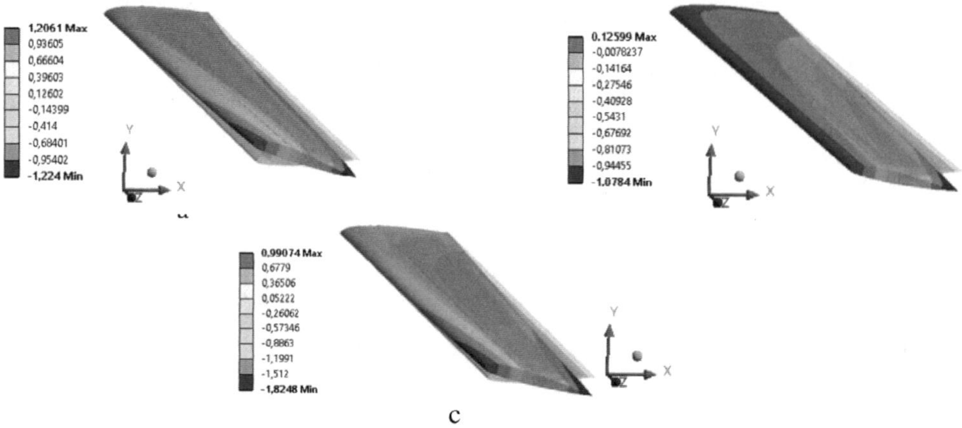

Figure 2. Fields of distribution of the blade segment displacements: a - GFRP and CFRP; b - CFRP; c - GFRP.

An analysis of the axial displacement fields U_y revealed that the maximum torsion angle was obtained for the GFRP helicopter blade segment. Table 2 presents the value of the helicopter blade segment torsion angles.

Table 2. Results of numerical simulation.

№	Spar material	Sheathing material	Torsion angle, °
1	CFRP	CFRP	0.57
2	CFRP	GFRP	1.15
3	GFRP	GFRP	1.33

It was found that the use of GFRP in the design of the blade segment provides the greatest torsion angles. The difference between the values of the torsion angles of the GFRP and CFRP blade segment is 57.14%. At the same time, the blade segment with the spar made of CFRP exhibits minimal longitudinal deformations. Thus, as a material, for the manufacture of a controlled helicopter blade segment with embedded piezoelectric actuators, an equal-strength GFRP is proposed.

At the next stage of the work, a numerical study of the effect of reinforcement schemes on the mechanical behavior of the blade segment was carried out. Based on the results of numerical simulation, the fields of stresses and displacements for a blade segment with embedded piezoelectric actuators were obtained. Figure 3 shows the axial displacement distribution fields U_y of the segment of the controlled helicopter blade at a control voltage of 1000V for the three reinforcement schemes.

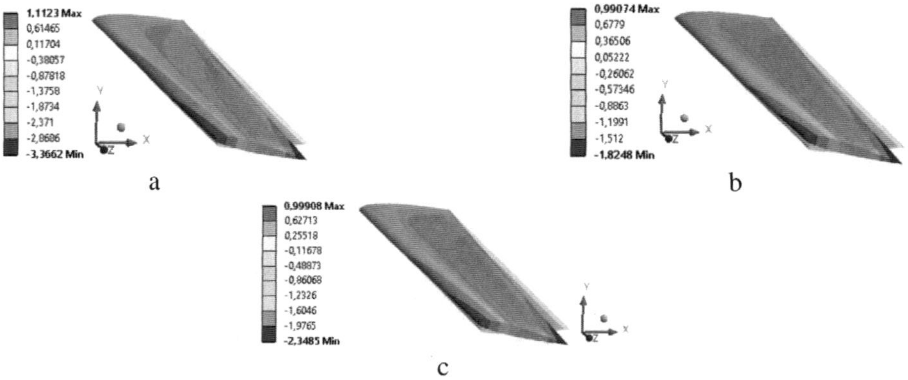

Figure 3. Fields of distribution of blade segment displacements with a reinforcement scheme: a - [0°/90°]; b - [45°/-45°]; c - [0°/45°/-45°/90°].

An analysis of the U_y axial displacements revealed that the largest torsion angles for the controlled helicopter blade segment sample from GFRP were observed for the reinforcement scheme [0°/90°]. Table 3 presents the results of numerical modeling of the blade segment mechanical behavior.

Table 3. The results of numerical simulation of the blade segment mechanical behavior.

№	Reinforcement scheme	Torsion angle, °
1	[0°/90°]	2.12
2	[45°/-45°]	1.333
3	[0°/45°/-45°/90°]	1.585

According to the results of numerical modeling, it was found that when using the reinforcement scheme [0°/90°], an increase in the torsion angle value is observed compared with the reinforcement schemes [45°/-45°] and [0°/45°/-45°/90°] by 37.12% and 25.24% respectively.

When designing a GFRP torsion structure, it is recommended to use the reinforcement scheme [0°/90°]. At the same time, in order to meet the requirements for this class of construction, it is necessary to assess the strength of the structure under operational loads at extreme angles of attack (12°).

To conduct a preliminary strength assessment, numerical calculations of the stress strain state of the segment were performed under the action of operational loads (lifting force Y + = 147.1 N, for an angle of attack of 12°). According to the results of numerical modeling of mechanical behavior under the action of operational loads, the distribution of stress and displacement fields was obtained for a blade segment with embedded piezoelectric actuators.

Analysis of the distribution of stress fields in the directions of layers reinforcing along the base and the weft in the coordinate system of the structure showed that they reach the greatest value in the region where the blade segment was attached. The concentration zones of these stresses were located in the area of attachment of the metal grip on the upper part of the blade segment Sheathing. Moreover, the maximum values of these stresses occurred on the surface of the outer layer of the blade Sheathing. The maximum normal tensile stresses along the base were σ_{11} = 78.581 MPa, along the weft σ_{22} = 54.276 MPa, and the maximum compressive stresses were σ_{11} = -74.155 MPa and σ_{22} = -61.161 MPa, respectively.

The maximum values of normal interlayer stresses were observed in the area of the leading edge between the first and second layer of the blade segment Sheathing. The compressive stresses σ_{33} reached an absolute value of 34.375 MPa, and the tensile stresses σ_{33} reached a value of 10.165 MPa. The maximum values of the interlayer tangential stresses τ_{13} were observed under a metal grip. The maximum tangential stresses were τ_{13} = 37 MPa.

For tangential stresses τ_{12} and τ_{23}, concentrations were observed near the metal fastening. The maximum value of tangential stresses in the reinforcement plane τ_{12} was 13.571 MPa. The maximum value of the interlayer tangential stresses τ_{23} was 19.548 MPa.

Assessment of the blades static strength was carried out according to the criterion of maximum stresses (1). The values of the static strength of the equal strength GFRP are given in table 4. The data presented was obtained experimentally, with the exception of the strength characteristics across the S_{33} layer, the estimated values of which are taken from [16].

Table 4. Values of permissible normal and tangential stresses.

S_{11}^+, MPa	S_{11}^- ,MPa	S_{22}^+, MPa	S_{22}^-, MPa	S_{33}^+, MPa	S_{33}^-, MPa	S_{12}, MPa	S_{13} MPa	S_{23} MPa
410	320	410	320	44	128	150	70	70

The assessment made it possible to determine the static strength reserves of the blade segment structure by the stresses in the layers. Margin of safety calculated by the ratio (2) are presented in table 5.

Table 5. Margin of safety coefficients.

k_{11}^{+}, MPa	k_{11}^{-}, MPa	k_{22}^{+}, MPa	k_{22}^{-}, MPa	k_{33}^{+}, MPa	k_{33}^{-}, MPa	k_{12}, MPa	k_{13}, MPa	k_{23}, MPa
5.22	4.32	7.55	5.23	4.33	3.72	11.05	1.89	3.58

Thus, at operational loads, the most dangerous in the design of the PCM blade segment are normal interlayer stresses. The margin of static strength of the GFRP blade segment is determined by these stresses and is 1.89.

4. Conclusion

As part of the research, a mathematical and numerical model for calculating the stress strain state of the controlled helicopter blade segment were developed. The choice of material and reinforcement scheme of the developed blade segment was made. It was revealed that when using the GFRP material and the reinforcement scheme [0°/90°], the largest torsion angle was observed, which amounted to 2.12°. Analysis of stresses in the layers of the helicopter blade segment revealed that the most dangerous, determining the margin of safety of the structure, are the normal interlayer stresses in the area. The margin of static strength of the GFRP blade segment at ultimate operating loads was 1.89.

Acknowledgments

The reported study was partially supported by the Government Perm Krai, research project No. C-26/797 dated 12/21/2017.

References

[1] Kovalovs A, Barkanov E and Gluhihs S 2007 *Journal of Physics: Conference Series* **93 1** 3-9
[2] Brudanov A M 2015 *Young Scientist* **104 24** 101-104
[3] Chen P and Chopra I *Smart Mate. Struct.* **5 1** 35-48
[4] Cesnik C E S and Shin S J *Smart Mater. Struct.* **10 1** 53-61
[5] Cesnik C E S and Shin S J 2001 *J. Solids Struct.* **38 10** 1765-89
[6] Chattopadhyay A, Lin Q and Gu H 2000 *AIAA J.***38 7** 1125-31
[7] Chattopadhyay A, Lin Q and Gu H 1999 *Composites: Pt B, Eng.* **30 6** 603-12
[8] Sekula M K, Wilbur M L and Yeager W T 1998 *Proc. Amer. Helicopter Soc. 4th Decennial Specialist Conf. Aeromech.* p 12.
[9] Sekula M K, Wilbur M L and Yeager W T 2005 *Proc. 61st Annual Forum Am. Helicopter Soc.* p 14
[10] Wilbur M L and Sekula M K 2005 *Proc. 61st Annual Forum Am. Helicopter Soc.* p 14
[11] Anoshkin A N, Rudakov M V, Straumit I S and Grinev M A 2011 *Proceedings of the Samara scientific center of the Russian academy of sciences* **13 1-2** 283-289
[12] Barkanov E, Anoshkin A N, Pisarev P V and Ashihmin V A 2019 *IOP Conf. Series: Materials Science and Engineering* **511** 012029
[13] Pisarev P V, Anoshkin A N and Ashikhmin V A 2018 *Scientific and Technical Bulletin of the Volga Region* **11** 144-147
[14] Makarova I S 2012 *Vestn. sam.gos tech. un-that. Ser. Phys.-mat. Science* **3 28** 191-195
[15] Pan'kov A A 2009 *Statistical mechanics of piezocomposites* (Perm. Gos. Tekn. Univ.) p 480
[16] Anoshkin A N, Tashkinov A A and Gritsevich A M 1997 *Mechanics of Composite Materials* **33 3** 255-262
[17] Anoshkin A N and Tashkinov A A 1997 *Mechanics of Composite Materials* **33 5** 449-454

IOP Conf. Series: Materials Science and Engineering **731** (2020) 012009 doi:10.1088/1757-899X/731/1/012009

The Influence of the Three-Layer Mouthguard on the Stress-Strain State of a Pair of Opposing Teeth

E S Avdeeva and A A Kamenskikh

Perm National Research Polytechnic University, Perm, Russia

avdeevaeliz@yandex.ru, anna_kamenskih@mail.ru

Abstract. The aim of the study was to consider the contact interaction of a pair of opposing teeth with and without taking into account the three-layer mouthguard. The strain-stress analysis of the biomechanical model was undertaken within the framework of the elastoplastic stress-strain theory. 7 geometric configurations of the mouthguard were considered within a wide range of functional loads which were varied from 50 to 500 N. We obtained the distribution of the stress and strain intensities in a pair of opposing teeth during contact interaction at a different level of physiological load. The dependences of the maximum level of stress intensity and the intensity of plastic deformations in the biomechanical model were established. Contact parameters were considered near the zone of dental occlusion.

1. Introduction

A biosocial cycle of the human body depends on such vital physiological processes as respiration and digestion, which are greatly influenced by the state of dentofacial system. In addition, teeth perform an aesthetic function, serve to process food mechanically and articulate speech [1, 2]. However, these functions may be impaired due to dental injuries that often occur in sport activities. For instance, in team kinds of sport, serious tooth injuries are the result of simultaneous impacts: falls, blows to the jaw, direct contact with a sports equipment. In power sports, thinning of enamel and tooth decay are caused by longstanding mechanical impact [2-4].

At the moment, the issues of biomechanical modeling aimed at preventing dental traumas in sports are of particular interest to scientists. Currently, one of the most effective ways to reduce the chance of injury or concussion of the brain, both during training and during competitions, is the usage of mouthguard or gumshield as a personal protective equipment [1, 4-7]. Due to the relevance of the usage of sports mouthguards, numerical modeling is required to describe the biomechanical behaviour of the protective devices and the materials, they are made of [8-9].

The aims of this paper are: a biomechanical analysis of the influence of the mouthguard material properties on the deformation behaviour of dental hard tissues; numerical modeling of contact interaction of a pair of opposing teeth taking into account protective devices of different geometric configurations within a wide range of physiological loads; a comprehensive interdisciplinary study of the stress-strain state (contact pressure) of a person's dentofacial system when using mouthguards.

2. Problem statement

The frictional contact interaction of a pair of opposing teeth was modelled with and without taking into account the three-layer mouthguard (figure 1).

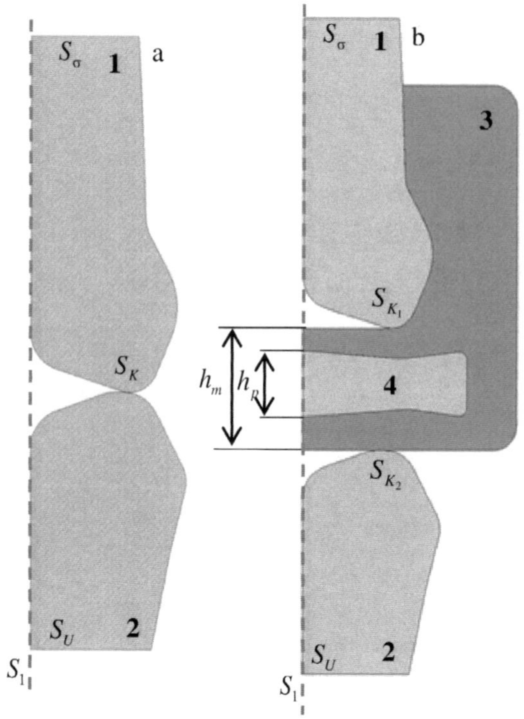

Figure 1. Computational schemes of contact interaction between a pair of teeth: a) without mouthguard; b) through three-layer mouthguard; 1 is maxillary tooth; 2 is mandibular tooth; 3-4 are three-layer mouthguard; 4 is a layer of A-silicone.

The influence of the geometrical configuration of the mouthguard on the stress-strain state of the biomechanical model is considered in this paper. Two types of the total mouthguard thickness are considered near the zone of dental occlusion $h_m = 6 \div 7$ mm and three types of the maximum interlayer thickness $h_p = 2 \div 4$ mm. The following types of geometrical configurations were regarded: 1 is $h_m = 6$ mm, $h_p = 2$ mm; 2 is $h_m = 6$ mm, $h_p = 3$ mm; 3 is $h_m = 6$ mm, $h_p = 4$ mm; 4 is $h_m = 7$ mm, $h_p = 2$ mm; 5 is $h_m = 7$ mm, $h_p = 3$ mm; 6 is $h_m = 7$ mm, $h_p = 4$ mm.

Mouthguard is a three-layer protective device: the main material is Eva (ethylene vinyl acetate); the interlayer is A-silicone. The material properties were obtained within a framework of an interdisciplinary study by the research team of PNIPU, PSMU and PSNIU [10]. Mechanical properties of materials are shown in a table 1.

Table 1. Mechanical properties of materials.

Material Name	**Young's modulus** E **(MPa)**	**Poisson's ratio** ν
Tooth material (tooth enamel)	$80.4 \cdot 10^3$	0.3
EVA	17.3	0.46
A-silicone	0.3	0.49

It was shown in the series of experiments that the Eva material exhibits elastoplastic properties, thus the dependence $\sigma - \varepsilon$ was plotted in [11]. To describe the behavior of the Eva material, the elastoplastic stress-strain theory is chosen. The mathematical statement is described in [11] and includes: equilibrium equations, physical and geometric relations, contact boundary conditions. The solution was realized in a plane statement with a constant thickness of the biomechanical model of 8 mm.

The mathematical problem statement (1)-(4) is supplemented by boundary conditions and symmetry conditions: a constant load is applied at the boundary S_σ and varies from 50 to 500 N

IOP Conf. Series: Materials Science and Engineering **731** (2020) 012009 doi:10.1088/1757-899X/731/1/012009

(indentation force); bending is prohibited; vertical displacements are prohibited at the boundary S_U; symmetry conditions are applied at the boundary S_1.

3. Results

The solution of the problem of the contact interaction of a pair of teeth through a three-layer mouthguard was implemented in the ANSYS Mechanical. An analysis of the convergence of the results of the numerical solution of the problem from the degree of discretization of the system was implemented earlier in [12]. It was established that a finite element mesh with a gradient with increased density of elements to contact zones gives an optimal solution to the problem in terms of accuracy and counting time. The maximum element size is 2.50E-04 m, the minimum is 6.25E-05 m. The finite element decomposition of biomechanical nodes, considered in this paper, was performed similarly to the previously selected grid.

The analysis of the influence of the thickness of the mouthguard and the interlayer of A-silicone on the stress-strain state of the node as a whole and the contact zone in particular was implemented. As a result, there were obtained: the fields of the distribution of stress intensity in biomechanical nodes and the intensity of total and plastic deformations in the mouthguard; the maximum intensity-stresses plots for the hard tissues of teeth and the intensity of plastic deformations-indentation force plots for the mouthguard. The parameters of contact interaction near the zone of dental occlusion were analyzed.

The field of stress intensity distribution for $F = 300$ N is presented in figure 2 taking into account the 6th variant of the protective dental device with $h_m = 7$ mm, $h_p = 4$ mm.

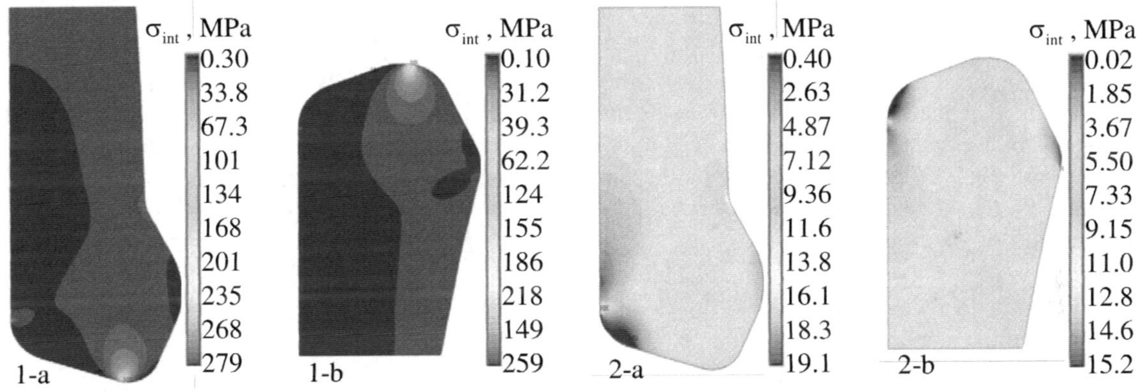

Figure 2. Stress intensity distribution in the teeth and mouthguard at indentation force 300 N: 1, 2 are computational scheme 1, 2 respectively; a is the maxillary tooth; b is the mandibular tooth.

In comparison with the computational scheme without mouthguard, the maximum stress intensity decreased by approximately 15 and 17 times for the maxillary and mandibular teeth, respectively, in the computational scheme with mouthguard. In addition, stress intensity fields were redistributed while considering contact interaction through the mouthguard. There were no local zones of maximum stresses near the zone of dental occlusion, the distribution of the stress intensity level became more uniform. The stress intensity in the mouthguard was minimal, which may be due to the physico-mechanical properties of the material or elastoplastic deformation of the mouthguard. The analysis of the influence of the geometric characteristics of the mouthguard on the deformation behaviour of the teeth was of a particular interest.

The dependence of the maximum level of stress intensity in hard tissues of teeth on the indentation force is shown in figure 3.

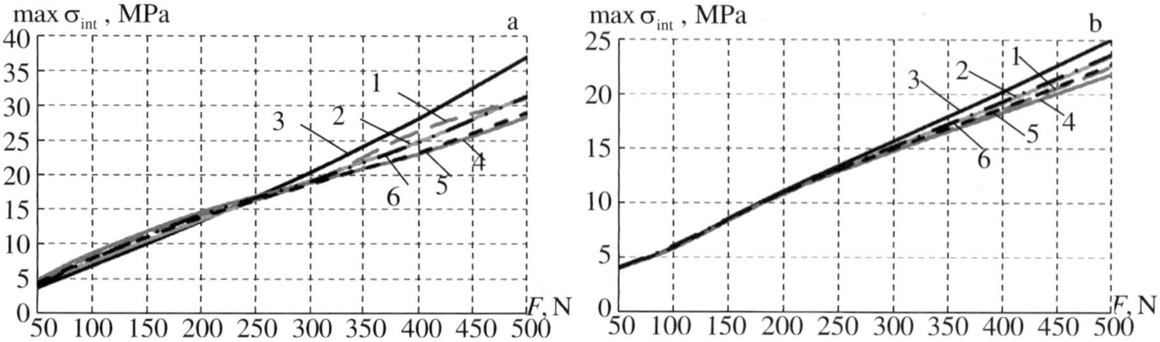

Figure 3. Maximum stress intensity-force plots for teeth contacting through the mouthguard: 1-6 are geometrical configurations of the mouthguard; a is the maxillary tooth; b is the mandibular tooth.

The total thickness of the mouthguard and the thickness of the interlayer have a significant impact on the deformation behaviour of the teeth. For a maximum indentation force of 500 N, the minimum values of the maximum stress intensity are 31.2 and 28.4 MPa with a mouthguard thickness of 6 and 7 mm, respectively. These values are observed for mouthguards with an interlayer thickness of 2 mm. Moreover, the volume of the Eva material in the zone of dental occlusion is maximum, due to the elastoplastic properties, the material is better adapted to the geometry of the opposing teeth and thus affects the stress reduction in the hard tissues of the teeth. In the maxillary tooth, the use of a 7 mm thick mouthguard reduces stress intensity for more than 15.75 times, a 6 mm mouthguard shows a less significant indicator of a decrease in stress intensity level (approximately 14 times on average). The decrease in the maximum level of stress intensity in the mandibular tooth is more significant and averages approximately 18.25 and 19 times for mouthguards with 6 and 7 mm thickness, respectively.

The value of plastic deformation in the main material of the mouthguard is of the particular interest. Figure 4 shows the dependence of the maximum level of plastic deformation in the mouthguard on the indentation force.

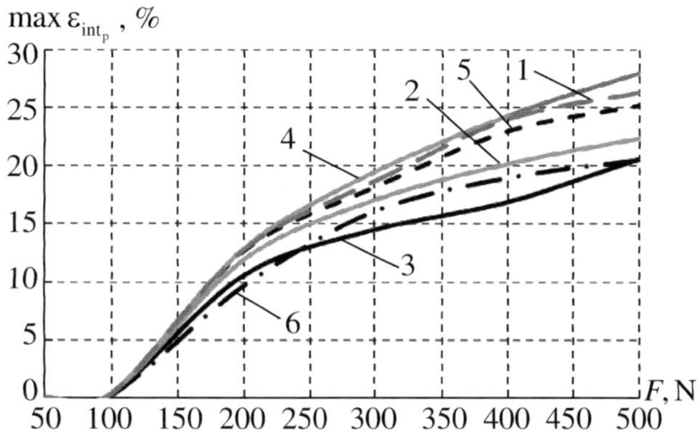

Figure 4. Maximum plastic strain intensity-force plots for the mouthguard: 1-6 are geometrical configurations of the mouthguard.

It can be noted that with an increase in the indentation force, the maximum level of plastic deformations increases according to a nonlinear law. Its maximum value is 26.3% with a maximum indentation force of 500 N. The maximum value of plastic deformations is observed in the model taking into account the 4 variants of the prosthetic design with $h_m = 7$ mm, $h_p = 2$ mm, which has the maximum decrease in the level of stress intensity in hard tissues of teeth.

Consider the contact parameters in the zone of dental occlusion during frictional contact interaction of opposing teeth through the three-layer mouthguard. Figure 5 shows histograms of the dependence of the maximum level of contact pressure and contact tangential stress on the indentation force. The distribution of contact parameters is shown for the biomechanical model for the 6th variant of the protective device.

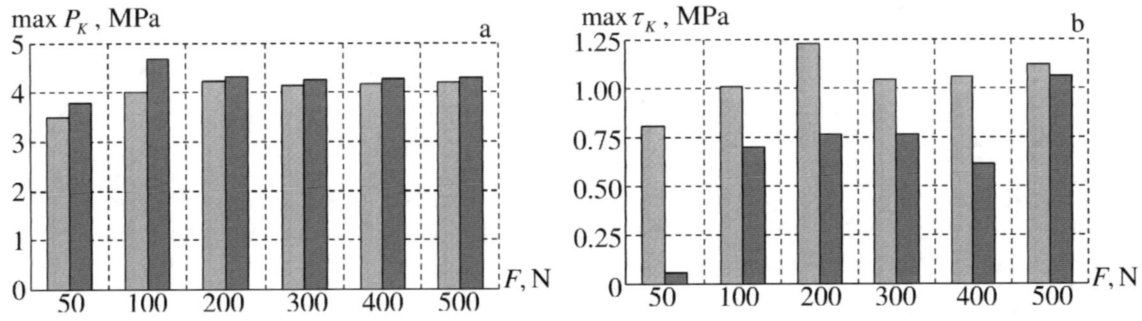

Figure 5. Dependence of the maximum level of contact pressure (a) and contact tangential stress (b) on the indentation force in the zone of dental occlusion during frictional contact interaction of opposing teeth through the three-layer mouthguard: light grey is maxillary tooth, dark grey is mandibular tooth.

The maximum contact pressure is greater in the mandibular tooth, and the maximum contact shear stress is greater in the maxillary tooth (figure 5). In addition, it can be noted that the contact shear stress is 4 or more times less than the level of contact pressure and is more susceptible to the geometric configuration of the mouthguard. With an increase in the indentation force, the value of the maximum contact pressure practically does not change; the maximum tangential stress is more sensitive to the load level.

4. Conclusion

As part of the study, a numerical model of the problem of contact interaction of a pair of opposing teeth was constructed with and without taking into account the three-layer protective dental device made of ethylene vinyl acetate and A-silicone. The analysis of the deformation behaviour of the biomechanical contact node within a wide range of functional loads of 50-500 N is performed. The influence of the thickness of the mouthguard and the layer on the stress-strain state of the biomechanical node as a whole is studied. The distribution of stresses and deformations in a pair of teeth during contact interaction at different levels of physiological load is obtained. When analyzing the results of a series of numerical experiments, it was established:

- When a pair of teeth comes into contact through a prosthetic structure, the maximum level of contact pressure is on average 10 times lower than when a person's teeth and jaw are in contact without taking into account the prosthetic structure. Contact shear stress is 4 times less than contact pressure. The effect of elastoplastic deformation of the mouthguard on the nature of the distribution of contact pressure and contact shear stress is established.
- With the contact interaction of a pair of teeth through a prosthetic design, a drop in the maximum intensity of stresses in the hard tissues of the teeth is observed. On average, the maximum intensity of tension in a tooth from the upper dentition during contact through the mouth guard is 11-15 times less; for a tooth from the lower dentition is 11-19 times less.
- Influence of the thickness of the layer of individually adapted mouthguards on the deformation behaviour of hard tooth tissues: the greatest decrease in the level of stress intensity is observed when the thickness of the mouthguard is not 2 mm, since during contact deformation the volume of the main mouthguard material in the tooth closure zone is maximal and adjusts to the geometry of the tooth pair during plastic material flow; with an increase in the thickness of the interlayer, the decrease in the

level of stress is less than that of a mouthguard with an interlayer of 2 mm by 3 and 10% with an interlayer thickness of 3 and 4 mm, respectively.

- The effect of the mouthguard thickness on the deformation behaviour of hard tooth tissues: a decrease in the maximum level of stress intensity in the hard tissues of mouthguards with a thickness of 7 mm is more by an average of 10 and 4% for teeth of the upper and lower dentition, respectively.
- With an increase in load, the area of plastic deformations in the mouthguard increases, while at a load of 500 N, the maximum level of intensity of plastic deformations does not exceed 20%. The effect of plastic deformation on the contact zone is manifested in the nature of the distribution of contact pressure and contact shear stress.

Acknowledgments

This work was financially supported by the Government of Perm Krai and Russian Foundation of Basic Research (project № 17-48-590411 r_a).

References

[1] Astashina N B, Kazakov S V, Ozhgikhina E S and Ozhgikhin Uy G 2014 *Actual problems in dentistry* **3** 34-37
[2] Giglio G D 2015 *J. of Am. College of Prosthodontists* **10** 1-4
[3] Zakaryan G A 2016 *Symbol of Sci.* **9** 140-146
[4] Kopetsky I S and Vasilyev Yu L 2016 *Dental Forum* **3** 32-34
[5] Knapik J J, Stephen W M, Robyn B L, Darakjy S S, Jones S B, Mitchener T A, DelaCruz G G and Jones B H 2007 *Sports Med.* **37** 118-144
[6] Retinskiy B V 2016 *Success of Mod. Sci. and Educ.* **8** 195-198
[7] Shemonaev V I, Klimova T N, Stepanov V A and Klimova N N 2016 *Mod Problems of Sci. and Educ.* **4** 82
[8] Lokhov V A, Astashina N B, Sergeeva E S, Merzlyakov A F and Baydarov A A 2017 *Russ. J. of Biomech.* **2** 211-218
[9] Chapman P J 1989 *Aust. Dental J.* **34** 36-43
[10] Lokhov V A, Kuchumov A G, Merzlykov A F, Astashina N B, Ozhgikhina E S and Tropin V A 2015 *Russ. J. of Biomech.* **4** 409-420
[11] Kamenskih A, Astashina N B, Lesnikova Y, Sergeeva E and Kuchumov A G 2018 *Ser. on Biomech.* **1** 3-15
[12] Kamenskikh A A, Ustjugova T N and Kuchumov A G 2018 *J. of Phys.: Conf. Ser.* **1129** 012014

IOP Conf. Series: Materials Science and Engineering **731** (2020) 012010 doi:10.1088/1757-899X/731/1/012010

Friction Induced Structural Changes of Mangalloy Observed by Synchrotron X-Ray Diffraction and Metallographic Analysis

K I Emurlaev, A. A. Ruktuev, A E Karmanova, V G Burov, A A Nikulina and I A Bataev

Department of Mechanical Engineering and Technologies,
Novosibirsk State Technical University, 20 Prospekt K. Marksa, Novosibirsk

emurlaev@corp.nstu.ru

Abstract. The effect of friction impact on structure and properties of mangalloy (Hadfield steel) was investigated. By means of metallographic analysis coupled with synchrotron X-ray diffraction in *operando* regime, the microstructural changes in specimen' subsurface layer during pin-on-disk experiment were studied. Analysis of phase composition showed that formation of martensite side by side with austenite has a cyclic behavior. Besides, it was shown that overall hardening is associated with a set of structural transformations. Finally, the failure of surface layer was observed.

1. Introduction

To date a lot of different steel grades for mold casting are known. The high-manganese steels that are one of the most common among iron-carbon alloys designed for operation under high contact loads of particular importance [1]. The development of mangalloy by R. Hadfield over 100 years ago was a starting point for its investigation. The chemical composition of Hadfield steel is shown in table 1. Known, that such alloys are characterized by high wear resistance at high specific and impact loads as well as by high degree of plasticity. Meanwhile, application of high manganese steels is limited due to extremely poor machining, metal forming and relatively low initial hardness. These alloys are mainly used for cast parts that do not undergo subsequent machining [2].

Table 1. Chemical composition of Hadfield steel.

C	Mn	Si	Cr	Ni	Cu	S	P	Fe
			less than					
0.90 - 1.40	11.50 - 15.00	0.30 - 1.00	1.0	1.0	0.30	0.050	0.120	balance

It's a well-known fact that high strength of austenitic steels is usually achieved by cold working instead of heat treatment. In doing so, one of strengthening mechanisms is a strain induced formation of α'- and ε-martensite as it was demonstrated in the first half of the 20th century [2]. Though, high degree of work hardening was also observed in the absence of these structural components that was confirmed by X-ray diffraction analysis. The mechanical twinning was found in such cases [2-5] and most of the studies were conducted using light and electronic (both scanning and transmission) microscopy.

Due to the wide spread of high manganese iron carbon alloys like Hadfield steel the mechanisms of their strengthening still require detailed consideration. The microstructural evolution under sliding friction are of special interest since it could combine cyclic loading, severe plastic deformation and heating. The non-destructive methods based on microscopy are typically used to study features of structural transformations. However, this approach is not always rational since an experimental sample subjected to certain number of friction cycles cannot be subjected to further tests after microscopic studies. Therefore, a new sample is required to study subsequent structural alterations caused by higher number of revolutions. This procedure, in turn, can lead to a mismatch in the test conditions. In addition, the interruption of the tribological process can lead to the removal of wear debris, cooling or heating of the studied object during sample preparation for metallographic observation that leads to a change in the structure of the subsurface layers, the thickness of which is often units of micrometers.
It is for that reason that methods of structural evolution characterization using synchrotron radiation is of particular interest [6]. High brilliance of radiation provides an opportunity to obtain hundreds of diffraction patterns per second in *operando* or *in situ* regimes. This paper presents the results of the dry sliding friction process of Hadfield steel obtained with friction tester with simultaneous use of synchrotron microdiffraction.

2. Experimental procedure

2.1. Description of the experimental scheme

The test was carried out according to the «pin-on-disc» friction geometry at the ID13 beamline of the European Synchrotron Radiation Facility (ESRF). A schematic representation of the experiment is shown in figure 1. The beam energy and the beam size were 13.9 keV and 2.8×1.3 μm respectively. The inclined radial surface of the disk was irradiated using synchrotron beam behind the pin at a distance of approximately 3 mm. The slope of 3.4° degrees contributed to an increase in the beam projection on the studied surface up to 2.8×22 μm. During friction diffracted beam was recorded using two-dimensional detector DECTRIS EIGER 4M with resolution 2070×2167 pixels and a maximum frequency 750 Hz. Diffraction ring patterns have been azimuthally integrated and analyzed using scripts developed in the *Python* programming language. Instrumental broadening was taken into account via α-Al_2O_3 standard. A detailed description of the experiment is presented in previous works [7, 8].

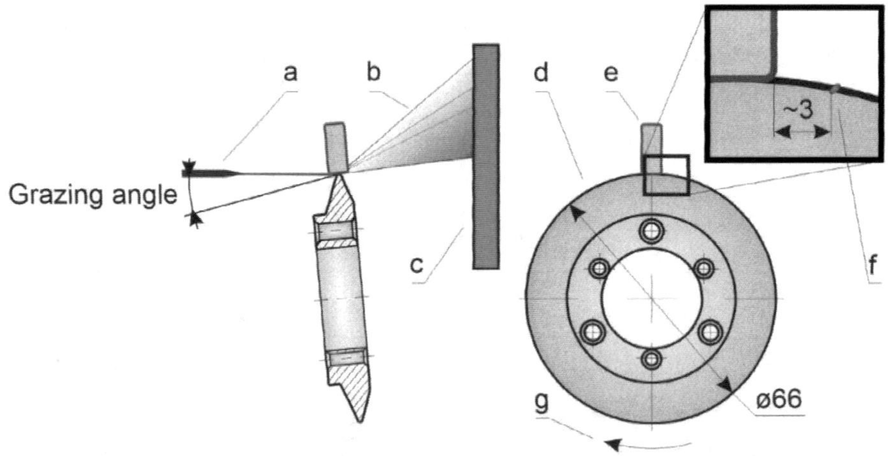

Figure 1. A scheme of a pin-on-disk experiment with simultaneous synchrotron X-ray diffraction: (a) – incident synchrotron beam; (b) – diffracted beam; (c) – 2D detector; (d) – disk-shape sample; (e) – WC-Co pin; (f) – trace of the incident beam; (g) – sliding direction.

The disc sample was made of high manganese austenitic Hadfield steel (elemental composition is shown in Table 2) via lathe machining. The radial surface of the disc was carefully polished using diamond pastes after which vacuum quenching was carried out with a temperature of 1000 °C in the oil. No traces of oxidation were observed on the surface of the sample after heat treatment. It was assumed that the sample preparation eliminated all the effects of previous machining and that any remaining structural imperfections, such as microstresses, dislocations, etc. were only a consequence of the heat treatment.

Table 2. Elemental composition of studied material.

C	Mn	Si	Cr	Ni	Cu	S	P	Fe
			less than					balance
1.060	12.750	0.79	0.070	0.062	0.255	0.005	0.045	

A pin was made from tungsten carbide-cobalt alloy. This material was chosen because of its high mechanical properties, in particular high wear resistance. During the experiment, the steel sample was pressed against a rigidly fixed pin. This approach allowed to keep the position of the disk surface unchanged when it wears out.

The experiment was implemented under dry friction conditions, the contact pressure was 160 MPa, and the angular velocity was 10 rpm.

2.2. Methods of material characterization

Upon the completion of the experiment the sample was cut using a wire electrical discharge machining along the direction of friction and perpendicular to the axis of rotation. Surface preparation was carried out by grinding on abrasive paper and polishing using diamond suspensions. The microstructure was investigated with scanning electron microscope (SEM) Carl Zeiss EVO50 after etching in a solution of nitric acid and water in a ratio of 1:9. Microhardness was measured by the Vickers method at a load of 0.098 N in three regions at different distances from the contact surface, each value was obtained by averaging 5 measurements. The elemental composition of the alloy was analyzed using optical emission spectrometer ARL 3460.

3. Results and discussion

3.1. The evolution of phase composition

The studies conducted using synchrotron X-ray diffraction have shown that structure of subsurface layers consist of austenite and martensite mixture until at least 100 friction cycles (figure 2). The presence of the ε-phase was observed only once after 30 cycles of friction.

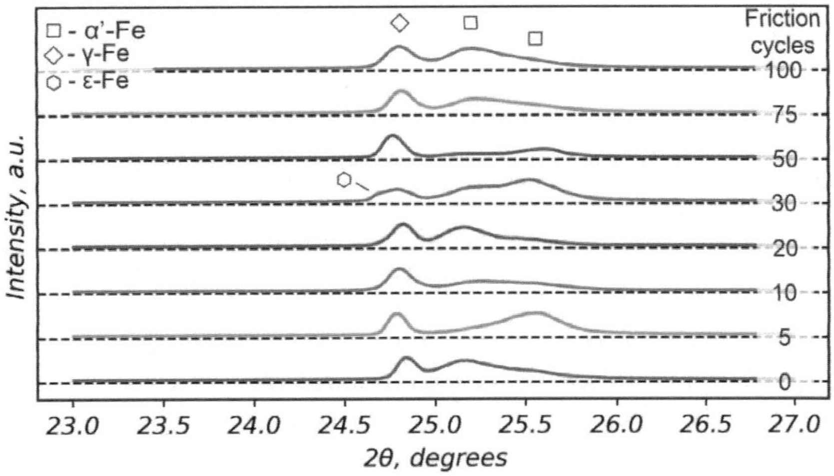

Figure 2. X-ray diffraction patterns obtained at different stages of friction with the use of synchrotron radiation.

IOP Conf. Series: Materials Science and Engineering **731** (2020) 012010 doi:10.1088/1757-899X/731/1/012010

An unexpected result is the presence of α'-martensite at the beginning of the experiment. Probably, its presence is due to the deformation of the material during quenching that could be caused by thermal stresses associated with the high cooling rate.

It may be noted that the alteration in the height of martensite peaks in dependence on number of revolutions has a cyclic behavior. It is also aware that the relative intensity of a single peak of a certain phase is associated with the atomic structure of the crystal and the volume fraction of this phase as well as significantly depends on the presence of crystallographic texture [9]. In addition, alteration of peak shape in particular the full width at half maximum (FWHM) is associated with subgrain refinement, lattice distortions and accumulation or annihilation of dislocations [10]. Based on the analysis of the obtained results including figure 3 we believe that the accumulation of damage of the subsurface layers that lead to the strain induced transformation of austenite into martensite, its subsequent deformation, failure and removal in the form of wear debris are observed from the outset. At the same time, «fresh» - lightly deformed areas of γ-iron are opened then the described process is repeated. So, for instance, upon ten-twenty friction cycles the relative intensity of supersaturated solution α' decreases compared to the diffraction maximum of austenite. Apparently, it is associated with fracture and removal of α'-martensite from the contact area and the appearance of fresh γ-phase regions on the sample surface. Further, the austenitic structure adapts to external loads that is expressed in dislocation glide as well as the rotation of the microvolumes of the specimen in which slip planes are oriented parallel to the sliding direction [11]. In the meantime, we can observe the increase in austenite' FWHM. And then after the 50-th revolution the fraction of austenite increases again to which indicates repeating the of the above described process.

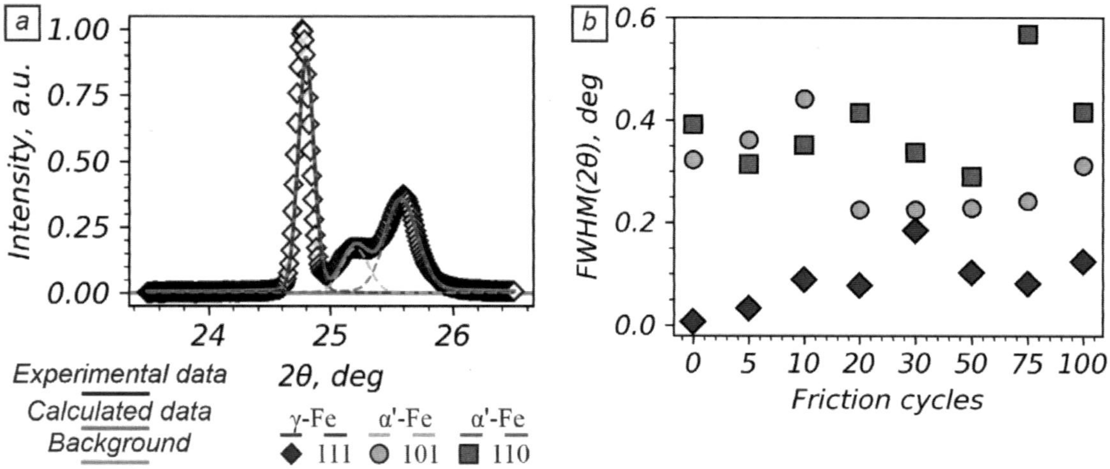

Figure 3. Diffraction line fitting (a) and alteration of FWHM in dependence on friction cycles (b).

3.2. The fracture of subsurface layer

Figure 4 shows the surface of a high manganese steel disc subjected to friction loading. The structure of the contact spot allows to conclude that severe plastic deformations occurs on the surface which contributed to the separation of the material and formation of wear debris involved in friction interaction [12]. Besides, it is possible to note the presence of microcracks, which also confirm the assumption of the destruction of the surface layer, leading to the removal of martensite from the contact region [11, 13].

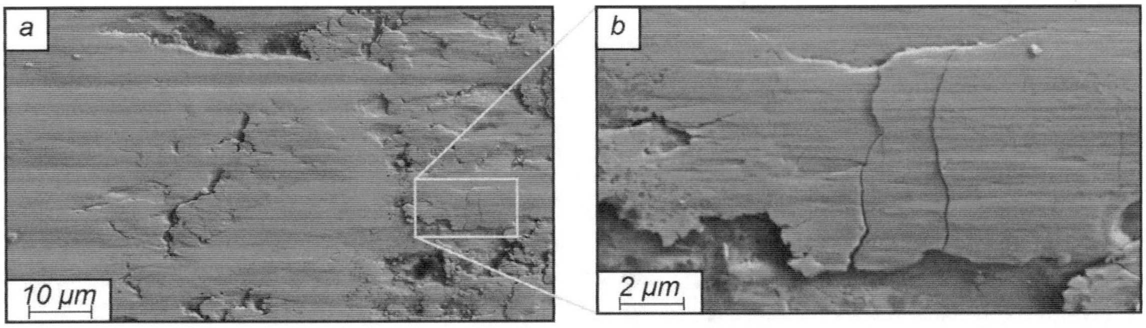

Figure 4. The disk surface after test.

The microstructure pictures obtained from the longitudinal section of the sample show that microcracks propagate in direction perpendicular to the friction surface (figure 5). This behavior is typical for delamination processes. The thickness of the surface layer, presumably consisting of a mixture of austenite, martensite, and wear products is approximately 1 μm. The area below is austenite grains with densely spaced slip bands. The distribution of microhardness from the contact surface at a distance from the surface of 10, 70 and 205 μm is 576.0±106.8, 458.2±45.2 and 396.0±48.8 *HV*, respectively. Thus, strengthening is associated with grain refinement, slip bands and dislocations as well as precipitation of strain induced martensite which probably has nanoscale size. Nevertheless, the hardening mechanism requires more detailed consideration, including the study of the fine structure of the material.

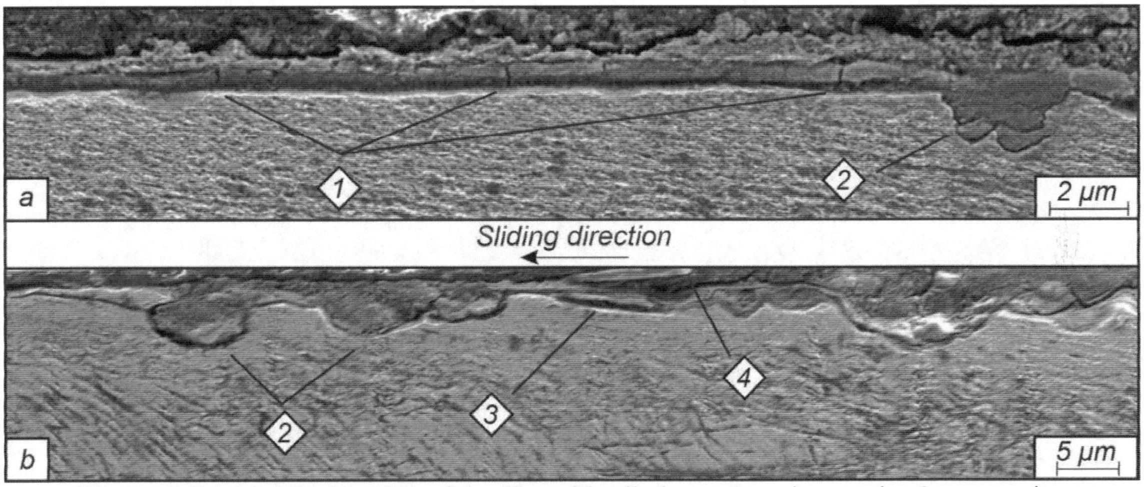

Figure 5. Longitudinal section of the disc with tribolayer: 1 – microcracks; 2 – wear pits; 3 –microcrack parallel to sliding direction; 4 – wear particle.

4. Conclusions

Based on this study, it can be concluded that the hardening of the deformed layer is achieved with a set of structural transformations and not limited to strain induced α'-martensite generation. Formation of slip bands, structure refinement and rotation of microvolumes also contribute to increasing of hardness. It is shown that the presence of martensite in the interaction region has a cyclic behavior due to the processes of deformation-induced transformation and delamination.

Acknowledgments

The reported study was funded by RFBR and Novosibirsk region according to the research project № 19-48-543022.

References

[1] Bouaziz O, Allain S, Scott C, Cugy P and Barbier D 2011 High Manganese Austenitic Twinning Induced Plasticity Steels: A Review of the Microstructure Properties Relationships *Curr. Opin. Solid State Mater. Sci.* **15** 141–68

[2] S Malinov L, E Malysheva I, Klimov E, Kukhar V and Balalayeva E 2019 Effect of Particular Combinations of Quenching, Tempering and Carburization on Abrasive Wear of Low-Carbon Manganese Steels with Metastable Austenite *Mater. Sci. Forum* **945** 574–8

[3] Lychagin D, Filippov A, Kolubaev E, Novitskaia O S, Chumlyakov Y and Kolubaev A 2018 Dry sliding of Hadfield steel single crystal oriented to deformation by slip and twinning: Deformation, wear, and acoustic emission characterization *Tribol. Int.* **119** 1–18

[4] Nikulina A, Smirnov A and Velikosel'skaya E 2013 Structural changes in Hadfield steel under cold deformation *J. Surf. Investig. X-ray, Synchrotron Neutron Tech.* **7**

[5] Astafurova E G, Tukeeva M S, Melnikov E V, Kretov Yu L, Nikulina A A and Velikosel'skaya A Yu 2013 The effect of aluminum on the regularities of structure refinement in single crystals of Hadfi eld steel under high-pressure torsion *Obrabotka Metallov / Metal Working and Material Science* **1** pp 58-62

[6] Yagi K, Kajita S, Izumi T, Koyamachi J, Tohyama M, Saito K and Sugimura J 2016 Simultaneous Synchrotron X-ray Diffraction, Near-Infrared, and Visible In Situ Observation of Scuffing Process of Steel in Sliding Contact *Tribol. Lett.* **61** 19

[7] Burov V, Bataev I and Smirnov A 2017 In-situ study of structural evolution of tribological materials using synchrotron radiation *MATEC Web Conf.* **129** 2024

[8] Bataev A, Burov V, Nikulina A, Bataev I, Lazurenko D, I. Popelukh A and Ivanov D 2018 A Novel Device for Quasi In Situ Studies of Materials Microstructure during Friction *Mater. Perform. Charact.* **7** 20170065

[9] K. De A, C. Murdock D, Mataya M, G. Speer J and Matlock D 2004 Quantitative Measurement of Deformation-Induced Martensite in 304 Stainless Steel by X-Ray Diffraction *Scr. Mater.* **50** 1445–9

[10] Ungár T, Gubicza J, Ribárik G, Borbely A, Unga Âr T, Riba Ârik G and Borbe Âly A 2001 Crystallite Size Distribution and Dislocation Structure Determined by Diffraction Profile Analysis: Principles and Practical Application to Cubic and Hexagonal Crystals *J. Appl. Cryst.* 298±310 J. **34** 298–310

[11] Rigney D, Chen L H, Naylor M G S and Rosenfield A R 1984 Wear processes in sliding systems *Wear* **100** 195–219

[12] Tarassov S Y and Kolubaev A V. 1999 Effect of friction on subsurface layer microstructure in austenitic and martensitic steels *Wear* **231** 228–34

[13] Suh N 1973 The delamination theory of wear *Wear* **25** 111–24

Influence of a Delamination Defect Location on the Operating Life of a Multi-Layered Composite Sample

A N Anoshkin, P V Pisarev and D A Ermakov

Perm National Research Polytechnic University, Perm, Russia

E-mail: pisarev85@live.ru

Abstract. In the framework of this research, the authors proposed an algorithm for predicting the residual life of structures made of polymer composite materials (PCM) using the structural-phenomenological model. To assess the residual life of PCM structures under cyclic loading, it was proposed to use the approach of explicit description of the adhesive layers destruction process. In accordance with the developed algorithm, in the process of cyclic loading in the adhesive layers, the level of damage was recorded; upon reaching a critical damage level, the adhesive layers destruction was simulated. The algorithm was tested on sample structures with a defect in the form of delamination and without defects. A numerical study of the delamination defect location influence at the onset of fatigue failure of model structures was carried out.

1. Introduction

Composite materials are widely used in the construction of aircraft structures. The use of composites can reduce weight, improve manufacturability and, in some cases, reduce costs. During operation, parts and components of the aircraft are subjected to various kinds of dynamic loads. As a result of prolonged exposure to dynamic loads in composite structures, damage accumulates, which, in the future, can lead to fatigue failure [1]. Therefore, an important point in the design and development of PCM aircraft structures is forecasting of structure operating life, which should be based on new mathematical models of the mechanics of composite materials.

A review of publications on the problem of modeling the fibrous composite materials fatigue strength [2-5] made it possible to obtain an extensive list of developed models - about 60. These models can be divided into three main groups: a) models based on the Palmgren-Miner hypothesis, b) phenomenological models that include two subclasses: residual strength models and residual stiffness models, c) - successive damage models, including models based on fatigue life, damage increase models and damage accumulation models.

Models based on the Palmgren-Miner hypothesis introduce a damage parameter without physical interpretation using the effects of fatigue damage accumulation. Most of these models were originally used for metallic materials. The nonlinear dependence of the long-term strength on such parameters as the load sequence and the level of fatigue stresses is taken into account. They include models such as the Macro and Starkey model [6] (modified linear law of static damage summation, 1954), Owen and Howe (based on observations of the cracks development in fiberglass, 1972), Hashin and Rotem (1978.) and others. Models based on the linear Miner rule still play a very important role in the study of composite materials fatigue strength under the action of a variable amplitude load.

Of particular interest are phenomenological models that relate the process of material fatigue damage to physically measurable quantities such as residual stiffness or strength. For most phenomenological models, the material parameters should be experimentally determined for each configuration of the layers package (laminate). The aim of the residual strength models is to assess the drop in static strength after a certain number of load cycles [7–9]. Residual stiffness models were developed on the basis of non-destructive residual stiffness measurements during long-term loading, which is the main advantage over residual strength models.

In the framework of this work, a mathematical formulation of the problem and an algorithm for the sample resource numerical prediction using the structural-phenomenological model were proposed. A numerical study of the delamination type defect location influence on the values of the model structures minimum operating time to failure was carried out.

2. Numerical model

The object of this study was a multilayer composite sample (Figure 1) made by autoclave molding. The sample consisted of 33 carbon fiber reinforcing layers with a reinforcement scheme [0°/45°]. Lay up of the layers was carried out to the central layer with a reinforcement scheme 0. Thus, all even layers had an angle of [45°], uneven layers - [0°].The number of layers in the sample corresponds to the number of layers in the composite flange joints used in aircraft products. The geometric dimensions of the sample are as follows: length 170 mm, width 35 mm, height 6.5 mm.

Figure 1. Appearance of a multilayer composite sample.

To predict the operating life of a multilayer PCM sample, the algorithm presented in [10, 11] was used. The algorithm is based on the finite element method using kinetic equations with a scalar damage function.

In accordance with the developed algorithm, the level of damage in the adhesive layers during cyclic loading was recorded. Upon reaching a critical level of damage, the destruction of adhesive layers was modeled. In the destroyed elements of the adhesive layers, the elastic constants were assumed to be approximately equal to zero (multiplied by 10^{-6}), which corresponds to the interlayer destruction of the layered package. To describe the damage accumulated in an elementary volume in the vicinity of a certain material point in the structure, we used the scalar function of time $\psi(t)$. It is believed that the $\psi(t)$ function takes values on the interval [0,1]. Moreover, the $\psi=0$ value corresponds to the case when there are no damages, the $\psi=1$ value corresponds to the level of damage at which the destruction of the elementary volume occurs. It was assumed that at the initial time t=0 for all points of the construction under consideration the value was $\psi=0$. The $\psi=1$ state corresponds to such a moment of time or the number of loading cycles N when a certain strength criterion is violated, depending on the stress state at a given point σ_{ij} and the material constants $S_{ij}(N)$.

We write the kinetic equation for damage accumulation, which determines the $\psi(N, \sigma_{mn})$ value, in accordance with the linear rule for summing damage:

$$\psi(N, \sigma_{mn}) = \sum_{k=1}^{N} \frac{1}{N_b(\sigma_i^{(k)})}, \quad (1)$$

where N_b is the function of the number of cycles to failure, $\sigma_i^{(k)}$ is the stress intensity in the k-th finite element.

The loading is assumed to be cyclic, symmetrical with an amplitude corresponding modulo the maximum static load. The influence of dynamic effects were neglected, the stress and strain fields in the structure at amplitude loads were considered equal to the corresponding fields under static loads. The strength of the adhesive layer under cyclic loads was taken according to [6]. Based on the given experimental data, a two-link fatigue strength curve was constructed, approximated by the equations:

$$lg[N_b(\sigma_i)] = -0.0684 \cdot \sigma_i + 9.18, \quad \sigma_i < 42 \text{ MPa},$$
$$lg[N_b(\sigma_i)] = -0.0684 \cdot \sigma_i + 9.18, \quad \sigma_i < 42 \text{ MPa}. \quad (2)$$

For each layer in each finite element, we obtain a nonlinear equation for N_b. After the solution the number of cycles before the destruction of each element N_{bj}^1 was determined. The minimum value N_{bj}^1 for all finite elements is the number of cycles until the first act of fatigue failure in the structure $N_{b\Sigma}^1$. After this, it was assumed that the corresponding element was destroyed and the elastic moduli were reduced in it.

The remaining elements received damage, which was calculated by the formula corresponding to the kinetic equation (1)

$$\psi_j^1 = \frac{N_{b\Sigma}^1}{N_{bj}^1}. \quad (3)$$

For the previous amplitude external load, a new stress-strain state of the structure with one damaged element q was calculated. Next, we determined the number of cycles before the destruction of each N_{bj}^2 element at the current stresses σ_{ij}. Failure in this case will occur in the element where the damage, taking into account the value already accumulated in the previous step (ψ_j^Σ), will be equal to unity:

$$\psi_j^\Sigma + \psi_j^2 = 1. \quad (4)$$

Damage at the current (second) step was determined by the formula:

$$\psi_j^2 = \frac{N_{b\Sigma}^{2add}}{N_{bj}^2}, \quad (5)$$

where N_j^{2add} - the additional number of operating cycles in the second step. Substituting (3) into (4), we calculate N_j^{2add} for each element, the minimum value of which is an additional operating time of the structure before the second act of destruction in the corresponding element. The total number of structure life cycles was calculated by the formula

$$N_\Sigma^2 = \min_j(N_j^{2\,add}) + N_\Sigma^1. \quad (6)$$

The tensor of elastic modules for the destroyed element was reduced, and for the remaining unresolved elements, the accumulated damage was calculated

$$\psi_j^\Sigma = \psi_j^1 + \frac{\min_j(N_j^{2add})}{N_{bj}^2} \quad (7)$$

The next step begins with the calculation of the stress strain state for the structure with two damaged elements.

The calculation of a structure fatigue life using this algorithm can be carried out until the structure exhausts it's bearing capacity for amplitude load, determined by the static tensile strength, or by a critical decrease in the structure rigidity, or by the formation of a given critical value damage zone. The final N_Σ value is the adjusted value of the structure fatigue life compared to the estimate obtained by the weakest link criterion.

One of the common types of defects in the manufacture of layered composite materials is delamination. In this work, we researched the influence of the location of the defect in the form of delamination on the life of a multilayer composite sample. The defect was located in the adhesive

IOP Conf. Series: Materials Science and Engineering **731** (2020) 012011 doi:10.1088/1757-899X/731/1/012011

layer, between the reinforcing layers. Numerical forecasting of the operating life in the ANSYS Mechanical software package was considered using the software developed by the authors on APDL.

The problem was solved in an elastic three-dimensional formulation. The analogy of cantilever fixing was used. From one end at a distance of 35 mm, nodes were identified that were limited by the displacements in the direction of the X, Y, Z axes: $U_x = 0$, $U_y = 0$, $U_z = 0$. The amplitude alternating load $U_z = U_{z0}$ was set to the edge of the opposite end. The boundary conditions are shown in Figure 2.

Figure 2. Scheme of boundary conditions.

The "delamination" defect was modeled as a region with a length of 10 mm, a width of 15 mm, and a height equal to the adhesive layer thickness, with reduced elastic properties (close to zero) [12]. Figure 3 shows the finite element model and the location of the defect in the adhesive layer.

Figure 3. The location of the defect in the finite element model of a multilayer composite sample.

The parameters of the research were the distance of the defect location from the edge of the sample (l) and the adhesive layer number (t). The l parameter took the following values: 25, 35, 45, 55, 65, 75, 85 (mm). The numbers of adhesive layers (t) considered were: 1, 4, 7, 10, 13, 16, 17, 20, 23, 26, 29, 32.

3. Numerical simulation results

At the first stage, a research was made of the defect location influence at the onset of fatigue failure and stress strain state. As a result of the numerical calculations, the values of the maximum stress intensities (σ_i^{max}) (Table 1) and the minimum number of run-in to failure (N_b^{min}) at the first loading step (Table 2) were obtained for the examined adhesive layers and defect locations, where t is the number of the adhesive layer, l is the distance from the defect fixing.

Table 1. Values of maximum stress intensities (σ_i^{max}, MPa) in the adhesive layers.

		Distance of the defect location from the edge of the sample (l)						
		25	35	45	55	65	75	85
Adhesive layer number (t)	1	25.8	26.24	25.8	25.8	25.8	25.8	25.8
	4	25.8	26.51	25.79	25.8	25.8	25.8	25.8
	7	25.8	30.85	31.22	29.17	28.19	27.6	27.16
	10	25.8	34.53	36.45	34.53	33.65	33.25	33.01
	13	25.8	36.84	39.99	38.25	37.4	37.09	36.95
	16	25.8	37.68	41.37	39.74	38.9	38.63	38.53
	17	25.8	37.68	41.37	39.74	38.9	38.63	38.53
	20	25.8	36.84	39.99	38.25	37.4	37.09	36.95
	23	25.8	34.53	36.45	34.53	33.65	33.25	33.01
	26	25.8	30.85	31.22	29.17	28.19	27.6	27.16
	29	25.8	26.51	25.79	25.8	25.8	25.8	25.8
	32	25.8	26.24	25.8	25.8	25.8	25.8	25.8

Table 2. The minimum values of run-in ($N_b^{min} \cdot 10^6$) at the first loading step.

	Distance of the defect location from the edge of the sample (l)						
Adhesive layer number (t)	25	35	45	55	65	75	85
1	25.986	24.265	25.995	25.991	25.989	25.989	25.989
4	25.986	23.245	26.029	26.005	25.996	25.993	25.988
7	25.986	11.750	11.072	15.294	17.852	19.576	21.005
10	25.986	6.578	4.861	6.578	7.56	8.045	8.351
13	25.986	4.574	2.785	3.662	4.184	4.393	4.492
16	25.986	4.005	2.238	2.896	3.301	3.448	3.504
17	25.986	4.005	2.238	2.896	3.301	3.448	3.504
20	25.986	4.574	2.785	3.662	4.184	4.393	4.492
23	25.986	6.579	4.861	6.578	7.56	8.045	8.351
26	25.986	11.751	11.072	15.294	17.852	19.576	21.005
29	25.986	23.246	26.029	26.005	25.996	25.993	25.989
32	25.986	24.266	25.995	25.991	25.989	25.989	25.989

Based on the results of numerical calculations presented in Table 2, the dependences of the minimum operating time to failure at the first loading step on the defect location were constructed (Figure 4, a). According to the results presented in Table 1, the dependences of the maximum stress intensities in the adhesive layers (σ_i^{max}) on the delamination defect location (l) in a multilayer composite sample were constructed (Figure 4, b).

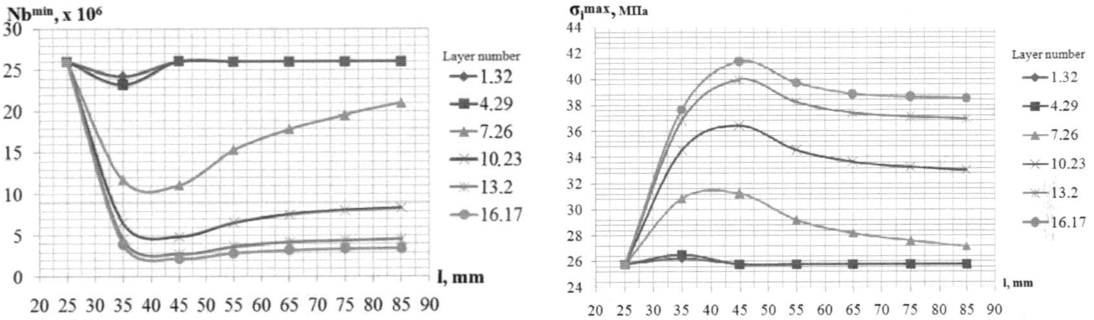

a b

Figure 4. The dependence of the minimum operating time to failure at the first loading step on the defect location (a), the dependence of the maximum stress intensities in the adhesive layers on the defect location (b).

An analysis of the obtained dependences revealed that the minimum run-in values until failure ($N_b^{min} = 2.237854 * 10^6$) are observed when the delamination defect is located in the central layers (t = 16, 17) of the multilayer sample at a distance l = 45 mm from the sample edge. Since when the defect is located 25 mm from the edge of the sample, the values of the minimum operating time to failure are equal, it can be argued that in the absence of defects in the sample under current loading conditions, $N_b^{min} = 25.98616 * 10^6$ cycles. It follows that at t = 16, 17 и l = 45, with a defect size of 10x15 mm, the life of the composite sample will be 11.61 times less than with a defect-free sample.

By analogy with the dependence presented in Figure 4, (a), the maximum stress intensities (σ_i^{max}) arise at t = 16, 17 and l = 45 mm σ_i^{max} at this location is 41.37 MPa, which is 1.6 times more than for a sample without a delamination defect.

4. Conclusion
As the main results obtained in the framework of this work, the following can be distinguished: an algorithm was developed for predicting the residual life of PCM structures using the structural and phenomenological model of composite material. In the framework of the developed algorithm, it was proposed to use the approach of the explicit description of adhesive layers destruction process to evaluate the PCM structures residual life under cyclic loading. In accordance with the developed algorithm, in the process of cyclic loading in the adhesive layers, the level of damage was recorded; upon reaching a critical level of damage, the adhesive layers destruction was modeled. The algorithm was tested on samples with and without defects. A numerical study of the influence of the delamination defect location at the time of the onset of fatigue failure and stress strain state was carried out. According to the results of the numerical study, it was found that for a sample without a defect under specified loading conditions, the values of the minimum operating time to failure amounted to $25.98616*10^6$ cycles. The minimum values of run-in until failure amounted to $2.237854 \cdot 10^6$ cycles and were detected when the delamination defect was the central layers. It was revealed that for a multilayer composite structure with cantilever bending, maximum stress intensities arise in the central layers. In this case, the location of maximum stress intensities and minimum run-in until failure is observed in the area of model structures cantilever fixing.

Acknowledgments
This work was supported by the Russian Foundation for Basic Research (project № 17-48-590129 p_a).

References
[1] Potrakhov N N, Anoshkin A N, Zuiko V Yu, Osokin V M, Pisarev P V and Pelenev K A 2017 *Vestnik PNIPU. Mechanics* **1** 118-133
[2] Van Paepegem W and Degrieck J 2001 *Compos Part A-appl S.* **32** 1433-1441
[3] Van Paepegem W and Degrieck J 2002 *Int. J. Fatigue* **24** 747-762
[4] Kaminski M, Laurin F, Maire J-F, Rakotoarisoa C and Hémon E 2015 *Aerospace Lab.* **9** 1-12
[5] Kennedy C R, Brádaigh C M and Leen S B 2013 *Composite Structures* **106** 201-210
[6] Macro S and Starkey W 1972 *Trans ASME***7** **6** 627-632
[7] Sevenois R D B and Van Paepegem W 2015 *Applied Mechanics Reviews* **6** **7**1-12
[8] Yang J N 1978 *Journal of Composite Materials.* **12** **1** 19-39
[9] Schaff J R and Davidson B D 1997 *Journal of Composite Materials* **31** **2**158-181
[10] Anoshkin A N, Pisarev P V and Ermakov D A 2018 *AIP Conference Proceedings* **2053** 040007
[11] Anoshkin A N, PisarevP V and Ermakov D A 2018 *AIP Conference Proceedings* **2051** 020017
[12] Pisarev P V, Anoshkin A N and Ermakov D A 2019 *IOP Conf. Series: Materials Science and Engineering* **511** 0120223

IOP Conf. Series: Materials Science and Engineering **731** (2020) 012012 doi:10.1088/1757-899X/731/1/012012

Influence of a Delamination Type Defect on the Operational Life of a Sound-Absorbing Structure with a Honeycomb Filler

A N Anoshkin, P V Pisarev and D A Ermakov

Perm National Research Polytechnic University, Perm, Russia

E-mail: pisarev85@live.ru

Abstract. In the framework of this research, the authors proposed an algorithm for predicting the residual life of sound-absorbing structures (SAS) from polymer composite materials (PCM) using the structural-phenomenological model. To assess the residual life of PCM structures under cyclic loading, it was proposed to use the approach of explicit description of the adhesive layer destruction process. The influence of the defect in the form of delamination on the honeycomb filler SAS operational life and stiffness was revealed. The description of the damage accumulation and destruction in the structure processes was obtained for the considered SAS under cyclic loading.

1. Introduction

Every year there is an expansion of the fields of polymer composite materials application. Composite materials are actively used in highly loaded units of aviation technology. Their use, in comparison with metal products, allows to reduce the mass, while maintaining the parameters of strength and rigidity of the structure [1-2]. To reduce the noise level that occurs during the operation of the aircraft, sound-absorbing structures (SASs) made of polymer composite materials are used. During operation, various types of dynamic loads act on the SAS, which, due to the susceptibility of polymer composite materials to the formation and accumulation of interlayer defects, can lead to fatigue failure. Therefore, the development of numerical methods for predicting the operational life and the process of composite structures destruction is an urgent task [3-6].

In the framework of this work, the effect of a defect in the form of a delamination on the operational life of a layered composite structure was researched. The object of the research was a single-layer SAS with honeycomb filler, shown in Figure 1. SAS consisted of two closed skins, between which there was a honeycomb filler. The geometric dimensions were as follows: length 200 mm, width 59.426 mm, height 16.089 mm. The casing consisted of two reinforcing layers of a woven equal strength prepreg with a reinforcement scheme [0; 0]. The honeycomb filler was made of fiberglass 1 mm thick and its height was 14.049 mm.

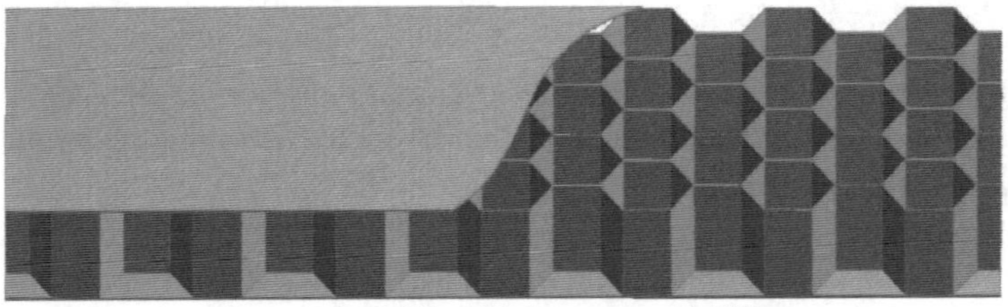

Figure 1. SAS with honeycomb filler.

2. Numerical model

To predict the honeycomb filler SAS operational life, the algorithm presented in [7-8] was used. The algorithm is based on the finite element method using kinetic equations with a scalar damage function. The multilayer composite structure under study is assumed in such a way that an isotropic adhesive layer is contained between the casing and the honeycomb filler. The thickness of the adhesive layer was taken equal to 10% of the thickness of the carbon fiber reinforcing layer. It was assumed that during the damage accumulation, the destruction of the structure will occur only in the adhesive layers.

Fatigue failure was modeled as a breakdown of the binder in the adhesive layers. In this way, the development of damage of the “delamination” type was simulated, which is a typical and characteristic type of these structures destruction, both under operational load and in laboratory tests [9]. The dependence of the strength of the binder on the level of cyclic loads was taken according to experimental data known in the literature [10]. A preliminary assessment of the residual life was carried out according to the weakest link criterion.

In the framework of computational experiments, a research was carried out of the effect of a defect in the form of a delamination on the honeycomb filler SAS residual life. A rectangular defect is located in the adhesive layer between the casing and the honeycomb filler (Figure 2). There is perfect contact between the casing, adhesive layer, defect and the filler. Numerical forecasting of the residual life was carried out in ANSYS Mechanical software package using the APDL program developed by the authors.

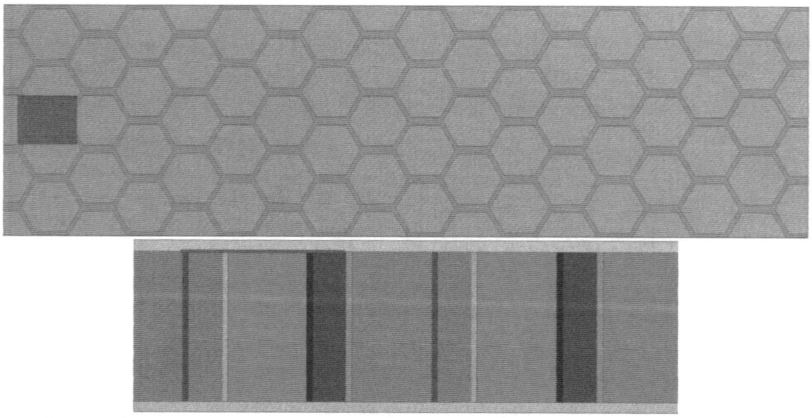

Figure 2. The location of the defect in the SAS with honeycomb filler.

The formulated problem of predicting the residual life of a SAS with a defect is solved numerically in a three-dimensional formulation. The boundary conditions were set as follows: cantilever fastening of one end of the SAS with a honeycomb filler; an alternating load is set to the opposite end, perpendicular to the honeycomb filler laying plane. The defect size was set in the form of a rectangular region with a size of 12.856x15 mm, the height was taken equal to the thickness of the adhesive layer.

As part of the computational experiments, 600 cyclic loading steps for each structure were considered. At each loading step in the numerical calculation the minimum operation life among all finite elements is determined. For the finite element with the minimal operation life, the reduction of elastic properties is performed.

According to the results of the numerical experiments, the dependences of the total number of running cycles (N_Σ) on the loading step (SL) for the SASs without the defect and with the defect were obtained. Dependencies are shown in Figure 3.

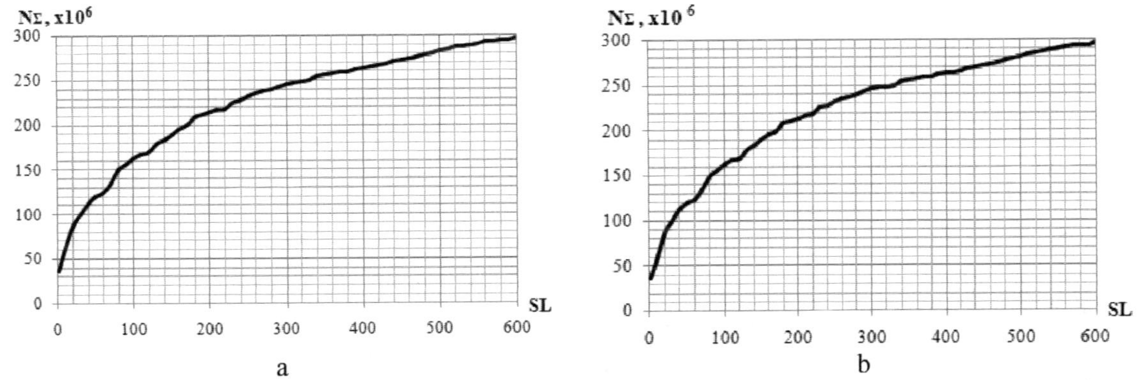

Figure 3. The dependence of the total operating life on the step of loading for the SAS without the defect (a) and the SAS with the defect (b).

An analysis of the results shows that in the first loading step, the minimum number of cycles to the first fatigue failure is $36.95 \cdot 10^6$. In the current research it was assumed that the minimum number of cycles before the failure determines the operation life of the structure.

The calculations for 600 loading steps that were carried out show that the dependence of the total operational life on the loading step is nonlinear. This pattern is caused by the reduction of material properties in the adhesive layers and the redistribution of the load.

We can conclude that this type of defect will slightly affect the total number of operation cycles of a honeycomb sound-absorbing structure sample.

During the cyclic loading of the SASs, the level of damage was recorded in the adhesive layers. Upon reaching the critical level of damage, the destruction of the adhesive layers was modeled through the reduction of the elastic characteristics in the damaged elements. At each loading step, a certain number of elements were destroyed. Figure 4, 5 shows the destroyed elements in the SASs with loading steps of 1, 100, 300, 600 for structures with and without the defect.

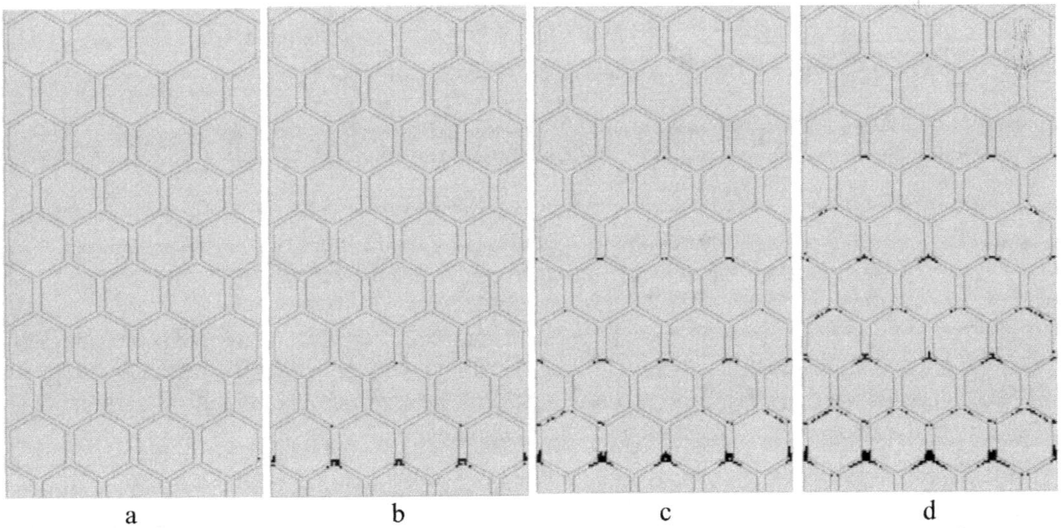

Figure 4. Areas of destruction of the SAS without defect at 10 (a), 100 (b), 300 (c) and 600 (d) loading steps (SL).

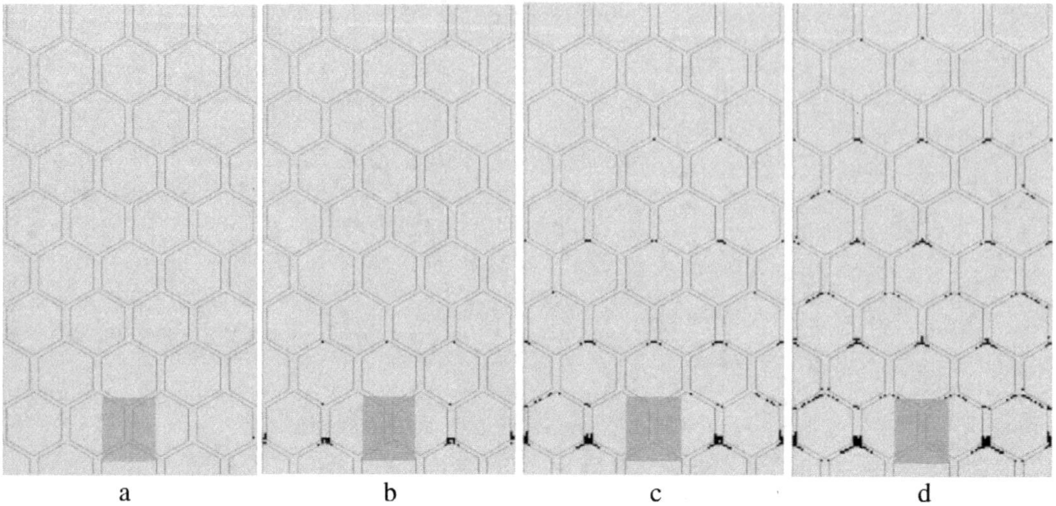

Figure 5. Destruction areas of the SAS with the defect at 10 (a), 100 (b), 300 (c) and 600 (d) loading steps (SL).

The analysis of the obtained results did not reveal (Figures 4 - 5, a - d) significant differences in the fatigue fracture of the adhesive layers of the SASs with and without the defect. The area of the destruction onset for both SASs is located directly in the zone of fixation (Figures 4, 5 a). Then there is a symmetrical distribution of the adhesive layers' destruction area (Figures 4, 5 b).

To analyze the reduction in the structural rigidity of the SAS structures with and without the defect, the dependences of maximum displacements on the number of loading steps are shown (Figure 6).

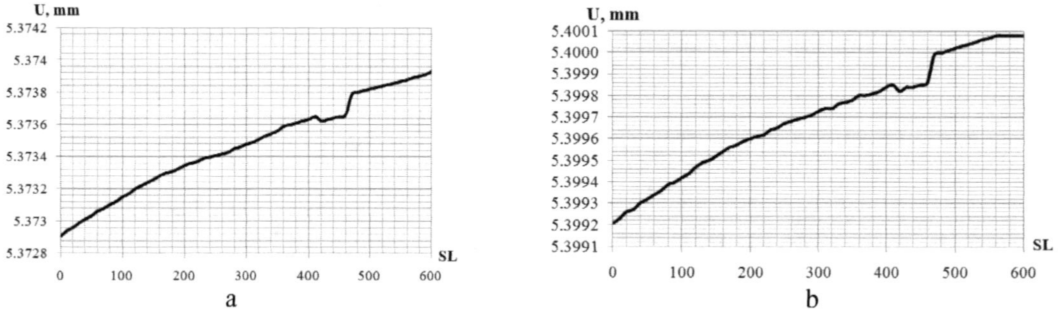

Figure 6. Dependences of maximum displacements (U) on the number of loading steps for the SAS without the defect (a), for the SAS with the defect (b).

From these graphs it can be seen that the greatest maximum displacements were observed for the structure with the defect, which is higher on average by 0.0263 mm than for the SAS without the defect. We can conclude that the presence of the defect in the form of delamination is insignificant due to a decrease in the structure rigidity during cyclic loading.

3. Conclusion

As a result of the research, the residual life was determined for the SAS with honeycomb filler. The effect of a defect in the form of delamination on the honeycomb filler SAS residual life, as well as the maximum displacements of the SAS without the defect and with the defect, were revealed. The description of the processes of damage accumulation and destruction in the profile of the considered structure under cyclic loading was obtained. The kinetic process of the SAS with a honeycomb filler destruction was simulated and described.

IOP Conf. Series: Materials Science and Engineering **731** (2020) 012012 doi:10.1088/1757-899X/731/1/012012

Acknowledgments
This work was supported by the Russian Science Foundation (project № 15-19-00259).

References

[1] Krause D 2016 *Composites Part B: Engineering* **87** 176-195
[2] Lueders C, Sinapius M and Krause D 2019 *International Journal of damage mechanics* **28** **10** 1523-1555
[3] Kennedy C R, Brádaigh C M Ó and Leen S B 2013 *Composite Structures* **106** 201–210
[4] Aeran A, Siriwardane S C, Mikkelsen O and Langen I 2017 *International Journal of Fatigue* **103** 327–341
[5] Benkabouche S, Guechichi H, Amrouche A and Benkhettab M 2015 *International Journal of Mechanical Sciences* **100** 180–194
[6] Zhou J, Huang H-Z, Barnhart M V, Huang G and Li Y-F 2019 *International Journal of Damage Mechanics* **0** **0** 1–16
[7] Anoshkin A N, Pisarev P V and Ermakov D A 2018 *AIP Conference Proceedings* **2051** 020017
[8] Pisarev P V, Anoshkin A N and Ermakov D A 2019 *IOP Conf. Series: Materials Science and Engineering* **511** 012023
[9] Bahrami M and Farrokhabadi A 2018 *ISME Conference* **6** **6** **1** 45-50
[10] Shokrieh M M, Esmkhani M and Haghighatkhah A R 2014 *Mechanics of Advanced Composite Structures* **1** 1–7

IOP Conf. Series: Materials Science and Engineering **731** (2020) 012013 doi:10.1088/1757-899X/731/1/012013

Nanostructuring of a Surface Layer as a Way to Improve the Mechanical Properties of Hypoeutectic Silumin

A A Kondratyuk[1], Yu F Ivanov[1,2], A A Klopotov[3,4], A M Ustinov[3], Yu A Abzaev[3], E A Petrikova[2], A D Teresov[2] and A S Tolkachev[1,2]

[1]National Research Tomsk Polytechnic University, 30, Lenin ave., Tomsk, 634050, Russia
[2]Institute of High Current Electronics SB RAS, 2/3, Akademichesky ave, Tomsk, 634055, Russia
[3]Tomsky State University of Architecture and Building, 2, Solyanaya sq., Tomsk, 634002, Russia
[4]National Research Tomsk State University, 36, Lenin ave., Tomsk, 634050, Russia

Email: alexkon@tpu.ru

Abstract. The irradiation of hypoeutectic silumin 383.1 with an intense pulsed electron beam in the melting mode and rapid crystallization of the surface layer has been performed. A multiphase submicron nanostructured surface layer with a thickness of up to 70 nm has been formed. Mechanical tests of the irradiated silumin samples in tensile experiments have been carried out. A significant increase in strength and plastic properties of silumin irradiated with an electron beam has been established. Features and patterns in the distribution of displacement fields in the deformation process in surface layers of the samples in realtime have been identified by digital image correlation method using the optical measuring system VIC-3D.

1. Introduction

Silumins are alloys of aluminum with silicon. Due to their low specific weight, relatively high specific strength, corrosion resistance, and good fluidity, silumins belong to cast alloys that are widely used in aircraft and shipbuilding, instrument making and construction, etc. [1]. A clear disadvantage of silumin is its high brittleness, conditioned upon the presence of coarse inclusions of silicon and second phase particles of the cast origin. A radical method of transforming the structure of silumin is high-speed heat treatment based on the use of intense pulsed electron beams of a submillisecond duration. In [2–4], it has been shown that the irradiation of silumin with an intense pulsed electron beam in the melting mode of the surface layer with a thickness of up to 1000 μm is accompanied by formation of a submicro nanocrystalline multiphase structure with high tribological and mechanical properties. Modification of the surface layer of silumin with an intense pulsed electron beam leads to formation of a gradient structure, the properties of which depend on the distance to the treatment surface. This circumstance significantly complicates the analysis of the deformation process of such materials. Speckle interferometry [5], implemented in [6–8] by using a three-dimensional digital optical system Vic-3D, is one of the areas of the analysis of the localization of plastic deformation of a material.

The purpose of the paper is to establish and analyze the distribution patterns of local plastic deformation foci under tensile conditions of flat samples of silumin 383.1 in the initial state and in the irradiated state with an intense pulsed electron beam.

2. Material and research technique

Silumin of the brand 383.1 was used as a research material. The irradiation of silumin was carried out on the "SOLO" installation [9] with the following parameters: the energy of accelerated electrons was 17 keV; the electron beam energy density was 25 J/cm^2; the irradiation pulse duration 150 μs; the number of pulses was 3; the pulse repetition rate was 0.3 s^{-1}. The irradiation was carried out in argon plasma at a residual pressure of 0.02 Pa. The elemental composition and the state of the defective substructure of the samples of cast silumin and silumin after electron beam irradiation have been studied using methods of scanning (Philips SEM 515) and transmission diffraction (JEM 2100F) electron microscopy [10–13].

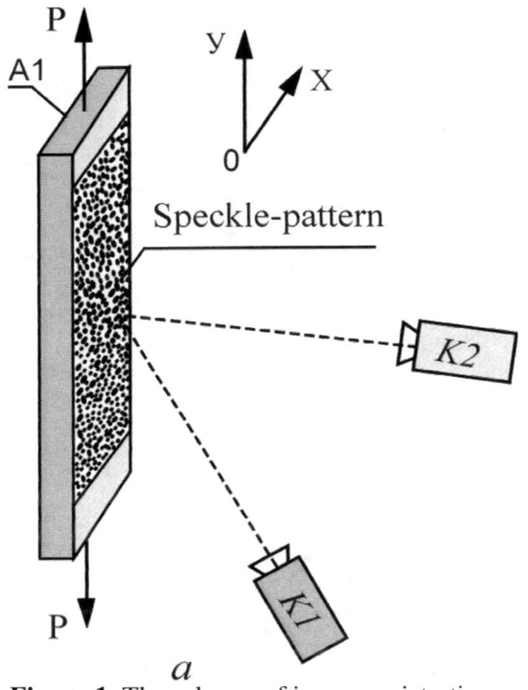

Figure 1. The scheme of image registration from a speckle-pattern on the lateral surface of the sample under tension; *K1* and *K2* are digital cameras; *P* is the applied load.

The tensile test was performed on an "INSTRON 3386" testing machine. The deformation rate was constant and was equal to 2 mm/min. The deformation distribution in surface layers of the sample under tension was obtained using the optical measuring system VIC-3D [14, 15]. The scheme of image registration on the surface of samples for obtaining digital stereoscopic images under tension is shown in Figure 1. Before tests, speckle structures were created on the surface of samples with contrasting fine white and black matte aerosol paints. Changes in displacement fields during the deformation process in real time were obtained based on processing of digital stereoscopic images using the VIC-3D optical measuring system. These displacement fields represent displacement projections of local regions of the sample surface along the OX axis (transverse deformation) and along the OY axis (longitudinal deformation). During further processing, using the software of the VIC-3D system, the displacements were converted to relative deformations (ε_{xx} is along the X axis, ε_{yy} is along the Y axis, ε_{xy} is the shear deformation). The deformation pattern of the sample surface is obtained by combining changes in microregions.

3. Research results and discussion

3.1. Data of electron microscopic studies

The structure of silumin 383.1 in the cast state is characterized by the presence of micropores, intermetallic inclusions, and micron-sized silicon (Figure 2). The presence of micropores and inclusions (especially lamellar shape), as a rule, contributes to a significant decrease in plastic properties of silumin [1].

The irradiation of silumin with an electron beam (25 J/cm^2, 150 μs, 3 pulses, 0.3 s^{-1}) leads to melting of the surface layer with a thickness of up to 70 μm (Figure 2, a, the layer is indicated by the arrow). The subsequent rapid crystallization allows forming a submicro-nanocrystalline structure of rapid cellular crystallization in the surface layer, the characteristic image of which is shown in Figure 3. The sizes of crystallization cells vary within 250-600 nm. The cells are separated by layers of the second phase, formed by particles with sizes of 40-80 nm. There are two types of crystallization cells in the modified layer. Firstly, the cells with a lamellar structure (Figure 3, a, the cells are indicated by arrows). The microdiffraction analysis shows that the plates are formed by silicon and are separated by layers of aluminum. The transverse dimensions of the plates and interlayers are 50-70 nm. Second, crystallization cells with a dislocation substructure, formed by an aluminum-based solid solution (Figure 4, a). Particles of the second phase of a round shape are revealed the volume of such cells (Figure 4, b). Particle sizes vary between 10-30 nm.

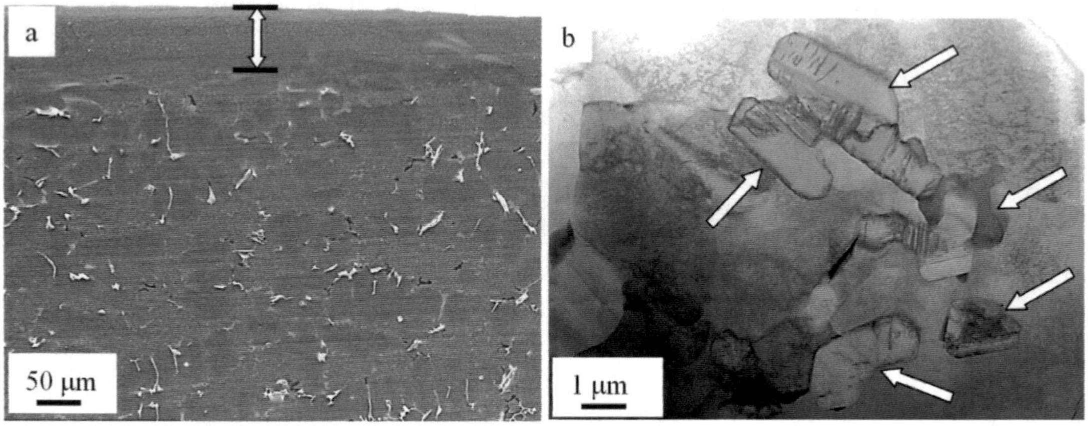

Figure 2. The structure of silumin383.1; a is the image obtained using methods of scanning electron microscopy (the arrow indicates the surface layer modified with an electron beam); b is the image obtained using methods of transmission electron microscopy (arrows indicate the inclusion of the second phase of the cast origin).

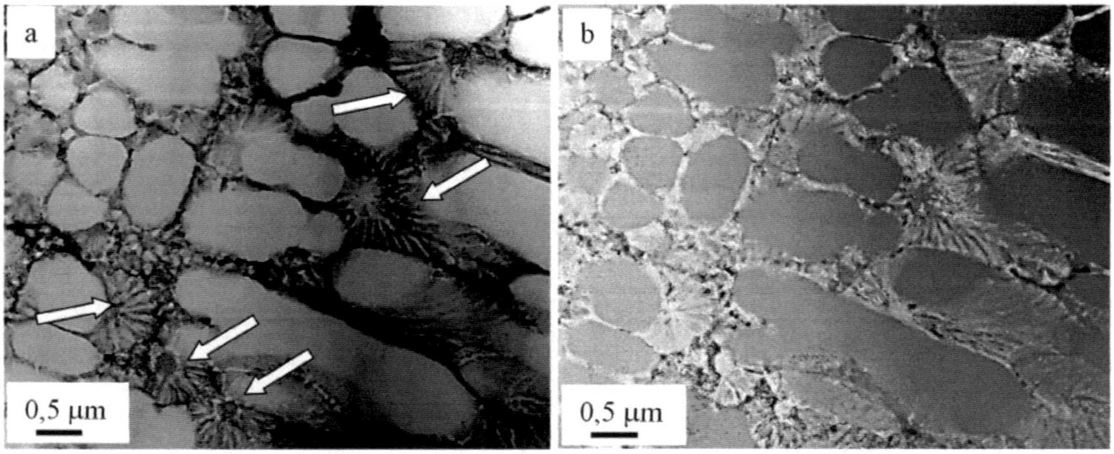

Figure 3. The structure formed in the surface layer of silumin 383.1 as a result of the irradiation with an intense pulsed electron beam; a is the bright field; b is the dark field; images have been obtained using methods of transmission electron microscopy (STEM method).

Thus, the irradiation of silumin 383.1 with an intense pulsed electron beam (25 J/cm^2, 150 μs, 3 pulses, 0.3 s^{-1}) allows forming a submicro-nanocrystalline multiphase structure in a relatively thin (≈70 μm) surface layer. It should be expected that such a transformation of the silumin structure will contribute to plasticization of the material.

3.2. Mechanical test data

The results of the tensile test of the non-irradiated silumin plate are shown in Figure 5. The reduced deformation curve is characteristic of materials with a high level of brittleness. The functional dependence $\sigma=f(\varepsilon)$ can be divided into 3 stages. This approach is standard in the analysis of deformation curves [16, 17].

Figure 4. The structure formed in the surface layer of silumin 383.1 as a result of the irradiation with an intense pulsed electron beam; on (b) the arrows indicate particles of the second phase.

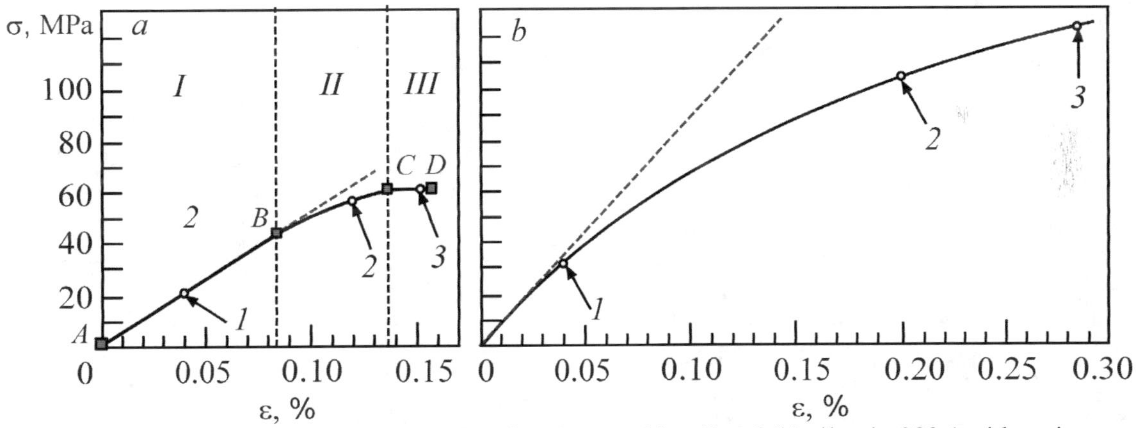

Figure 5. Deformation diagrams of non-irradiated (a) and irradiated (b) silumin 383.1 with an intense pulsed electron beam. The digits correspond to the position of deformation structure patterns on the curve (Figures 6 and 7). *AB* is the stage of elastic deformation (stage I); *BC* is the transitional stage (stage II); *CD* is the stage of deformation softening (stage III) and fracture.

Stage I is linear and reflects the elastic nature of silumin deformation. Stage II is transitional, followed by a very short stage of deformation softening (stage III). Points 1 and 2 on the deformation curve correspond to the distribution pattern of horizontal and vertical relative deformations on the silumin plate surface in Figure 6

Point 1 on the deformation curve (Figure 5, a) corresponds to deformation distribution patterns in horizontal (Figure 6, pattern 1_{XX}) and in vertical (Figure 6, pattern 1_{YY}) directions, where, in the central part of the sample, formation of a combined local deformation foci, consisting of two regions with positive and negative deformation values, is registered.

Point 2 on the deformation curve (Figure 5, a) corresponds to deformation distribution patterns, where, in the central part of the sample, it can be seen that a significant increase in deformation values has occurred in the plastic defamation foci.

The results of tensile tests of the irradiated sample of silumin 383.1 are shown in Figure 5б, b. It is noteworthy that in this diagram the functional dependence $\sigma = f(\varepsilon)$ can be represented as a parabolic function [18]. There are no deformation stages on the deformation curve of the irradiated sample. The presented deformation curves clearly show that the irradiation leads to hardening of silumin samples, and the destruction is observed at higher values than on the non-irradiated samples (Figure 5).

The analysis of distribution patterns on the irradiated samples has revealed a number of interesting patterns (Figure 7). Point 1 on the deformation curve (Figure 5, b) corresponds to the deformation distribution pattern in the vertical (Figure 7, pattern 1_{YY}) direction, where local regions with higher positive deformation values (ε_{YY}~0.13%) than the average deformation value on the sample (<ε>~+0.04%) have been formed in the upper right part of the sample. In the upper left part of the sample, symmetrically to the right part, local regions with negative values have been formed in the shape of layers (ε_{YY}~ – 0.05%) (Figure 7, pattern 1_{YY}). It is obvious that these local regions contain stress concentrators, which have appeared in the shape of local regions with higher deformation values.

An increase in the applied external stress leads to association of local deformation regions into larger plastic deformation foci (Figure 7, pattern 2_{YY}). The destruction of the sample has eventually taken place in regions where these plastic deformation foci have occured.

It should also be noted that local regions of compression ($\varepsilon_{XX} < 0$) and tension ($\varepsilon_{XX} > 0$) are observed in the distribution patterns of horizontal relative deformations on the surface of the irradiated sample (Figure 7).

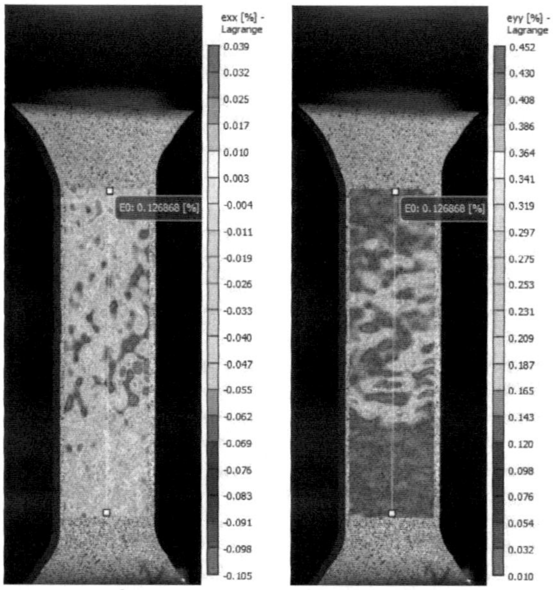

2_{XX} 2_{YY}

Figure 6. Distribution patterns of horizontal (digits with the XX index) and vertical (digits with the YY index) relative deformations on the surface of the non-irradiated sample. Digits 1 and 2 show in the diagram in Figure 5, a the corresponding deformation-stress states of the sample under test.

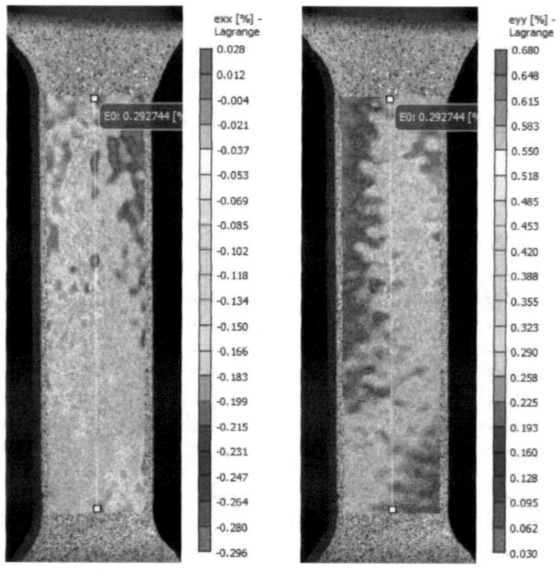

2_{XX} 2_{YY}

Figure 7. Distribution patterns of horizontal (digits with the XX index) and vertical (digits with the YY index) relative deformations on the surface of the irradiated sample. Digits 1 and 2 show in the diagram in Figure 5, b the corresponding deformation-stress state of the sample under test.

4. Conclusion

It has been shown that the irradiation of hypoeutectic silumin 383.1 with an intense pulsed electron beam in the melting mode and rapid crystallization of the surface layer leads to formation of a multiphase submicro-nanostructured surface layer with a thickness of up to 70 nm. Mechanical tests of irradiated silumin samples in tensile tests have been carried out. A significant increase in strength and plastic properties of silumin irradiated with an electron beam has been established. Features and patterns in the distribution of displacement fields in the deformation process in the surface layers of the samples in real time have been revealed by the digital image correlation method using the optical measuring system VIC-3D.

Acknowledgement

The research was financially supported by the RSF (project No. 19-79-10059).

References

Zolotarevskiy V S and Belov N A 2005 *Metallurgical science of cast aluminum alloys* (Moscow: MISiS)

[1] Laskovnev A P, Ivanov Yu F, Petrikova Ye A et al. 2013 *Modification of the structure and properties of eutectic silumin by electron-ion-plasma treatment* (Minsk: Belarus. Navuka)

[2] *Current trends in the modification of the structure and properties of materials* 2015 / Ed by N N Koval and V E Gromov (Tomsk: Publishing House NTL)

[3] *Electron-ion-plasma modification of the surface of non-ferrous metals and alloys* 2016 / Ed by N N Koval and Yu F Ivanov (Tomsk: NTL Publishing House)

[4] Jones R and Wykes C 1983 Hohgraphic and Speckle Interferometty (University Press. Cambridge)

[5] Ustinov A M, KIimenov V A, KIopotov A A, Abzaev Yu A and Ovcharenko V E 2018 Speckle Pattern of the Surface Layer of Titanium Alloy GRADE 2 and Steel 3723HR Samples under Deformation *Materials Science Forum* vol 938 pp 62-69

[6] Kopanitsa D G, Ustinov A M, Potekaev A I, Klopotov A A and Marchenko E S 2018 Russian Physics Journal vol 60 **9** pp 1577-1585

[7] Sutton M A, Orteu Jean-Jose, Schreier H W 2009 Image Correlation for Shape, Motion and Deformation Measurements. Basic Concepts, Theory and Applications *Springer Science, Business Media* 332 p

[8] Koval N N and Ivanov Yu F 2008 Nanostructurisation of surface of metal-ceramic and ceramic materials by pulsed electron beam processing *Izvestiya Vuzov. Fizika* **5** 60-70

[9] Thomas G, Goring M J 1983 *Transmission electron microscopy of materials* (Moscow: Nauka)

[10] Brandon D and Kaplan U 2004 *Microstructure of materials. Methods of investigation and control* (Moscow: Tekhnosfera)

[11] Utevskii L M 1973 *Diffraction electron microscopy in physical metallurgy* (Moscow: Metallurgia)

[12] Endrus K, Dyson D, Kyoyn S 1971 *Electron diffraction patterns and their interpretation* (Moscow: Mir)

[13] Manalo A, Sirimanna C, Karunasena W, McGarva L and Falzon P 2016 Construction and Building Materials vol 105 pp 365–376

[14] Vildeman V E, Lomakin E V, Tret'yakova T V and Tret'yakov M P 2017 Supercritical Deformation and Fracture of Bodies with Concentrators under Plane Stress State Condition. Mechanics of Solids vol 52 **5** pp 488-494

[15] Ustinov A M, KIimenov V A, KIopotov A A, Abzaev Yu A and Ovcharenko V E 2018 Speckle Pattern of the Surface Layer of Titanium Alloy GRADE 2 and Steel 3723HR Samples under Deformation *Materials Science Forum* vol 938 pp 62-69

[16] Zuev L B 2007 On the waves of plastic flow localization in pure metals and alloys *Ann. Phys.* vol **16** 4 pp 286-310

[17] Honeycombe R W K 1984 *The plastic deformation of metals* (Edward Arnold) 483 p

IOP Conf. Series: Materials Science and Engineering **731** (2020) 012014 doi:10.1088/1757-899X/731/1/012014

Electromagnetic Properties of Composite Materials Based on ABS Plastic with Carbon Nanotubes Obtained by the Additive Technology in the SHF and EHF Bands

G E Kuleshov[1,2], A V Badin[1,2] and M O Gering[2]

[1]Laboratory of radiophysical and optical methods of exploring the surrounding environment, National Research Tomsk State University, Tomsk, Russia
[2]Department of Radioelectronics, Faculty of Radiophysics, National Research Tomsk State University, Tomsk, Russia

E-mail: grigorij-kge@sibmail.com

Abstract. The results of study of the electromagnetic characteristics of polymer composite materials containing multiwall carbon nanotubes are presented. New radio filaments for 3D printing with MWCNTs have been created. Experimental samples were obtained from composite filaments by 3D printing. Also investigated is the ABS plastic brand Conductive with carbon fibers. Studies of electromagnetic characteristics were carried out in the SHF and EHF ranges. The frequency dependences of the complex permittivity for these materials are shown. The results show that ABS Conductive plastic has the best absorbing properties on microwave frequency, and composite plastics with MWCNTs of 2% or more on EHF.

1. Introduction

The rapidly developing 3D printing technologies today offer unlimited possibilities in the manufacture of structures and objects of any geometry for radioelectronic devices. Additive technologies have several advantages over traditional technologies, such as reducing the number of technological operations and time spent, waste-free use of source components, ease of development, optimization and modernization of products in engineering CAD-programs, high automation and manufacturability [1, 2]. Recently, it has been very common to manufacture composite materials based on plastics for 3D printing with the addition of fillers that effectively interact with electromagnetic radiation [3, 4]. Using of additive technology with radio filaments can significantly speed up and reduce the cost of the production of radio materials and protective coatings for end devices, and also makes it possible to create completely new elements and composite structures [5].

2. Materials and methods of measurement

Composition and structure of the materials was determined by X-ray diffraction analysis. Shimadzu XRD 6000 X-ray diffractometer and XRF-1800 wavelength dispersion X-ray fluorescence spectrometer were used. The samples of powders were studied

The study of the structure of the material was carried out using an optical microscope, as well as by transmission electron microscopy. Electron-microscopic measurements were carried out with a transmission electron microscope Philips CM 30 with electronic image recording.

2.1. Materials

Various polymeric filaments can act as the basis of a composite filament for 3D printing. The characteristics of the final sample can vary greatly from the choice of binder. Plastics are divided into several main types: ABS, PLA, HIPS, RUBBER, FLEX, CRYSTAL and etc.

In our case, the acrylonitrile-butadiene-styrene (ABS plastic) acted as the base polymer of the composite filament for 3D printing. Powders of multi-walled carbon nanotubes (MWCNTs) were used as filler [4]. They were obtained by catalytic gas-phase deposition of ethylene and contained more than 97.5% of the main phase with an average nanotube diameter d = 9.4 nm. TEM images of carbon materials are shown in Figure 1. MWCNT agglomerates in the form of tangles are observed.

Table 1. Characteristics of the MWCNTs.

Characteristics	Value of
Synthesis method	Catalytic gas-phase deposition of ethylene with FeCo catalysts
Average nanotubes diameter	9.4 nm (4 – 21 nm)
Filler content MWCNT, wt.%	more than 97.5
Impurities	particles FeCo/Al_2O_3
Ash content, wt.%	< 2.4
Metal impurities (Fe, Co, Ca), wt.%	< 1.7
External specific surface, m^2/g	320
Wall thickness of nanotubes	4.0 nm; 7-9 layers

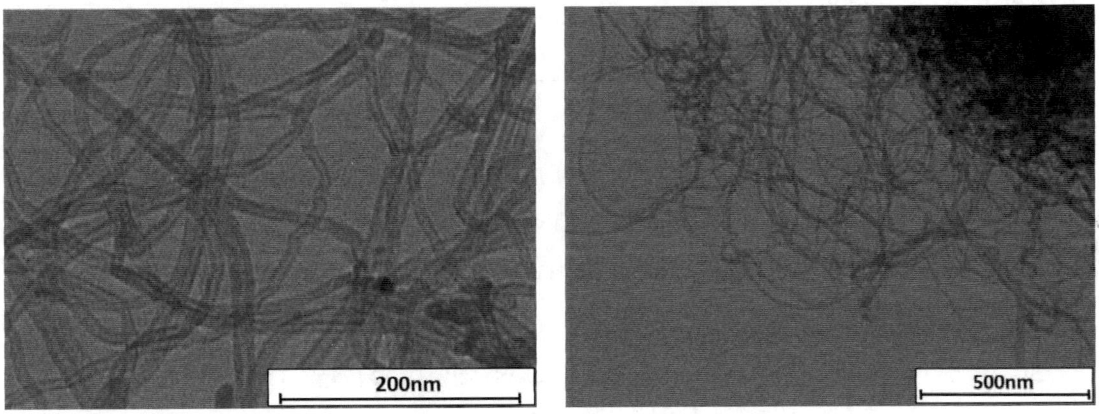

Figure 1. TEM images of the powders of MWCNT.

2.2. Obtaining experimental samples

Methodology for producing composite radiofilaments can be divided into several stages (Figure 2), each of which has an important role and will be described. For the production of composite plastic, the chemical method of mixing the components of the material was used. At the first stage, the required number of ABS granules (pieces of pure ABS) and filler were carefully weighed on an electronic balance with an accuracy of 0.001 g. At the next stage, the components are mixed, while the plastic used is previously dissolved to a homogeneous liquid state in acetone. The next step is the ultrasonic treatment of the unpolymerized mixture of ABS plastic with a filler. The processing was carried out on an ultrasonic device of the “Alena” series, model UZTA-0,1/28-O. The ultrasound power was 75 watts. This treatment allows to achieve greater uniformity and also breaks down the agglomerates of the filler, which is no less important. Since a dimethiketone solution was used to make the radiofilament, special attention had to be paid to the polymerization of the sample. After ultrasonic treatment, the liquid was placed in a container, where it solidified at room temperature. However, the evaporation of acetone vapor at room temperature occurs rather quickly, therefore, in order to avoid the appearance of empty cavities inside the sample, which will lead to anisotropy of properties, it is necessary to periodically stir the resulting mixture. After complete polymerization of the composite mixture, the resulting material was ground and hot extruded to create a filament. At least 20 iterations of extrusion are necessary to form a uniform filament structure. Resulting homogeneous filament was placed in the clip of 3D printer for further printing.

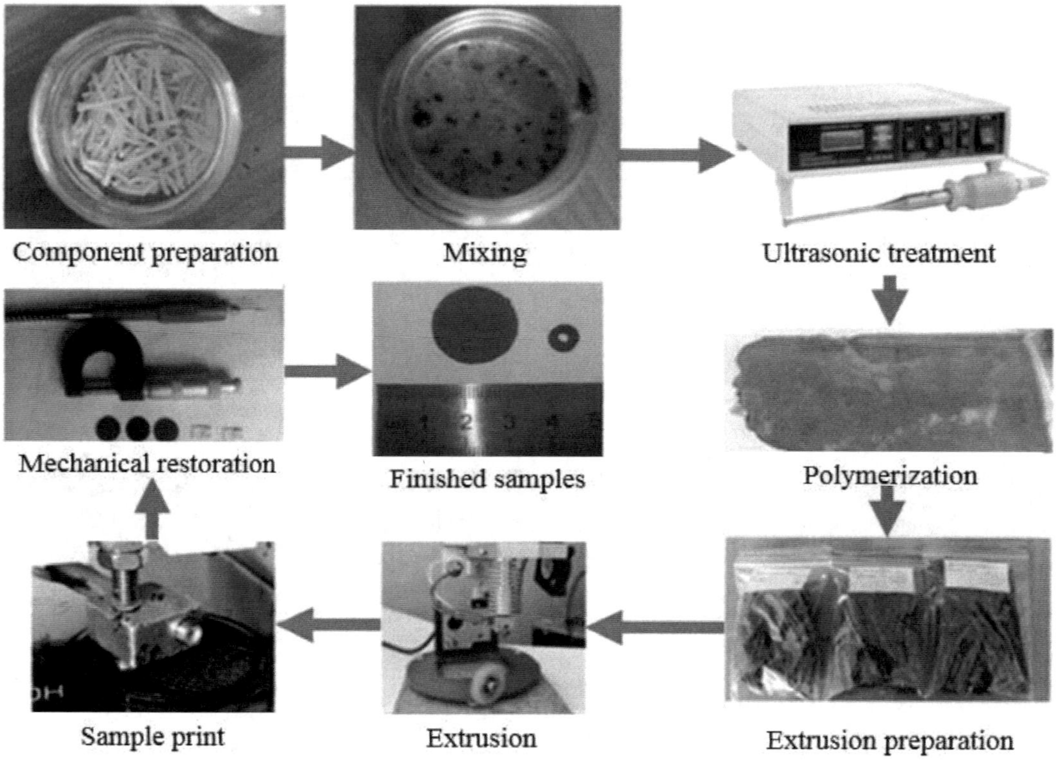

Figure 2. Production of samples of composite materials.

The diameter of the printer nozzle is 1 mm. As a result, experimental samples were printed in the form of coaxial washers (inner diameter of $d_{in} = 3$ mm, an external $d_{out.} = 7$ mm and a thickness of about $h = 2.25$ mm) for microwave research and in the form of flat cylinders for research in the UHF range. This allows you to cover the largest possible frequency band. The thickness of the fabricated samples is about 1 mm, and the diameter is about 2 cm.

Table 2 below shows the main parameters of the manufactured experimental samples for microwave measurements.

A total of twelve samples were made. Ten samples based on ABS plastic with the addition of 1, 2, 3, 4, and 5 wt.% MWCNTs (five for microwave measurements and five measurements for EHF), as well as an additional sample based on the industrially produced ABS-Conductive "Mellow" filament (China) containing carbon fibers (resistivity 10^3-10^5 Ohm /cm). EHF measurement samples are shown in Figure 3.

Table 2. Parameters of manufactured samples for a coaxial cell.

Sample No.	Binder, wt.%	Content of the filler, wt.%	Sample Thickness, mm
1	99 ABS	1 of MWCNT	2.26
2	98 ABS	2 of MWCNT	2.26
3	97 ABS	3 of MWCNT	2.25
4	96 ABS	4 of MWCNT	2.25
5	95 ABS	5 of MWCNT	2.24

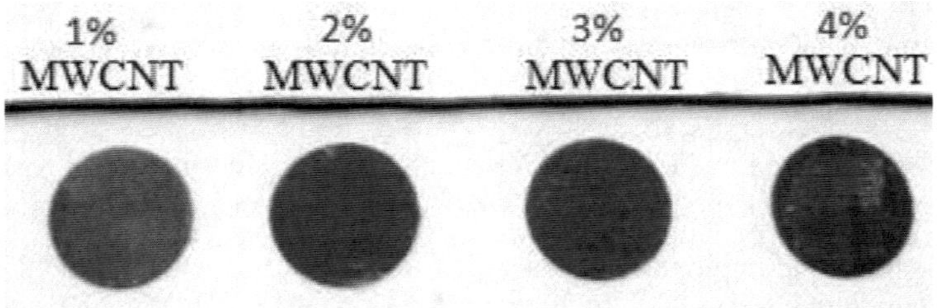

Figure 3. Manufactured and processed samples for EHF measurement.

2.3. Measuring equipment

The electromagnetic properties of the materials were measured by the waveguide method on a "Mikran" R4M-18 vector network analyzer in a coaxial cell according to the "transmission" schemes and on the STD-21 terahertz spectrometer.

Vector network analyzer allows you to measure *S*-parameters (transmission and reflection coefficients) with phase when placing the sample in the measuring cell. The frequency range is from 1 to 18 GHz. Further, using the obtained values of the transmission coefficients (*T*) and reflection (*R*) with their phase, one can find the spectra of complex magnetic and dielectric permittivity using the modified Becker – Jarvis method [6].

The electromagnetic characteristics of radio materials in the EHF range were studied using the free space method using the STD-21 terahertz spectrometer according to the standard "transmission" scheme in the frequency range 115–265 GHz. Backward wave lamps (BWT) were used as a radiation source, and a Galea cell was used as a receiver. The frequency dependences of the complex transmission coefficient were measured, which were further used in calculating the complex permittivity.

3. Results and discussion

Figure 4 shows the frequency dependences of the complex permittivity for samples based on composite radiofilaments.

It can be seen from the above dependences that with an increase in the concentration of carbon nanotubes in ABS plastic from 2 to 5 wt.%, An increase in the values of both real (from 3.5 to 7 rel.units) and imaginary (from 0.1 to 0, 7 rel.units) parts of the complex permittivity. The ABS Conductive plastic has ε' values comparable to ABS + 5% MWCNTs, but much larger ε'' values, which is probably due to conductivity losses, which decrease with increasing frequency. Thus, filaments with MWCNTs are suitable for creating shielding or narrow-band absorbing coatings at microwave, since they have high permittivity values.

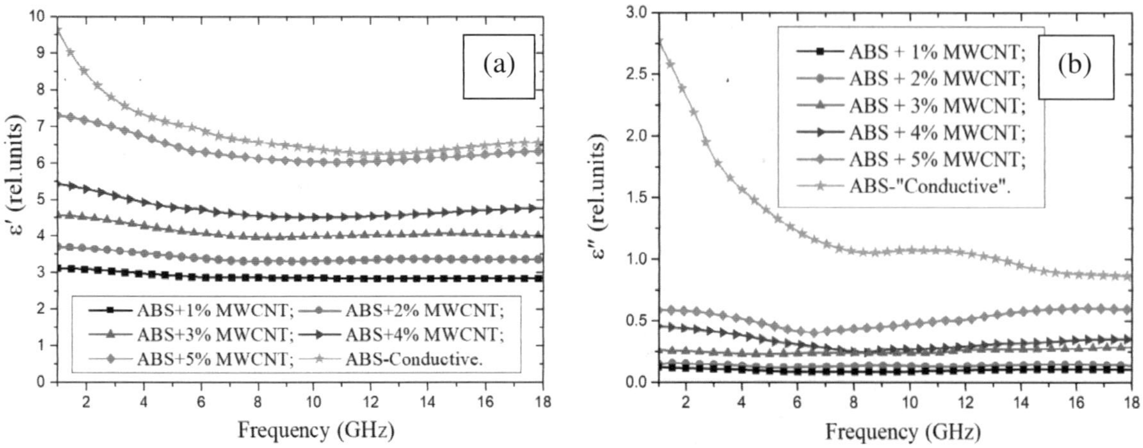

Figure 4. Frequency dependences of the real (a) and imaginary (b) parts of the complex permittivity for composite materials based on ABS plastic with carbon fillers.

Figures 5, 6 show the complex permittivity of the samples with MWCNTs, as well as the aforementioned ABS-Conductive Mellow sample. Figure 5 shows that in the considered frequency range, the ABS-Conductive «Mellow» sample has a linear frequency dependence of the real part of the complex permittivity. The samples with MWCNTs are characterized by a decrease in the real part of the permittivity with increasing frequency. It is also seen that for samples with a higher filler concentration, the decline is more avalanche. This is due to the polarization properties of the material and significant dielectric losses. In composites with a high concentration of MWCNTs, they are large (reach 2.4 rel.units at a concentration of 5 wt.%), and in plastic ABS Conductive ε" are small (0.9 rel.units), therefore, it has an almost linear frequency dependence. Figure 6 shows an almost linear frequency dependence of the imaginary part of the complex permittivity for all the samples studied.

For the ABS-Conductive sample, the imaginary part of permittivity takes values close to 0.8 rel. units, and for plastic samples with MWCNTs, dielectric losses linearly increase with increasing filler concentration. Already at a concentration of 2 wt.% MWCNT ε" is greater than that of ABS-Conductive, and at a concentration of 5 wt.% It reaches 2.4 rel.

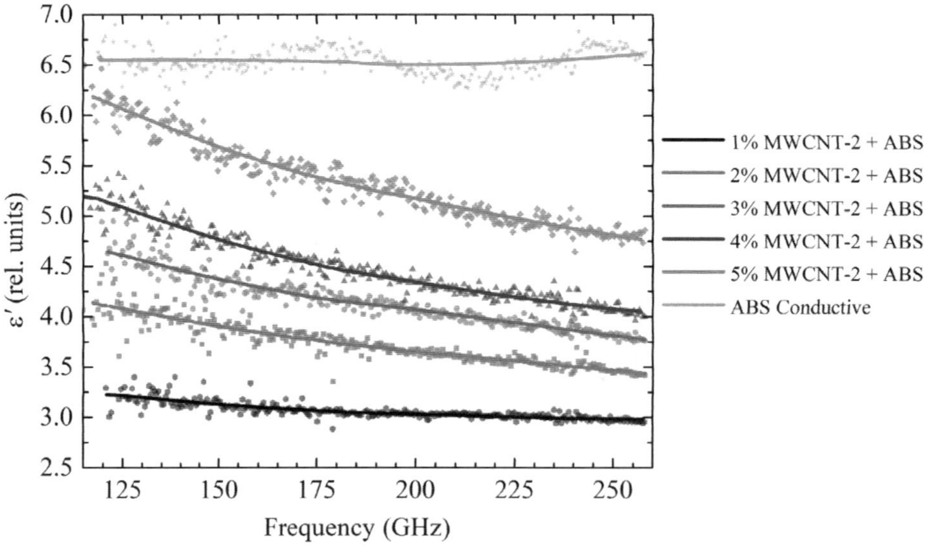

Figure 5. Frequency dependence of the real part of permittivity.

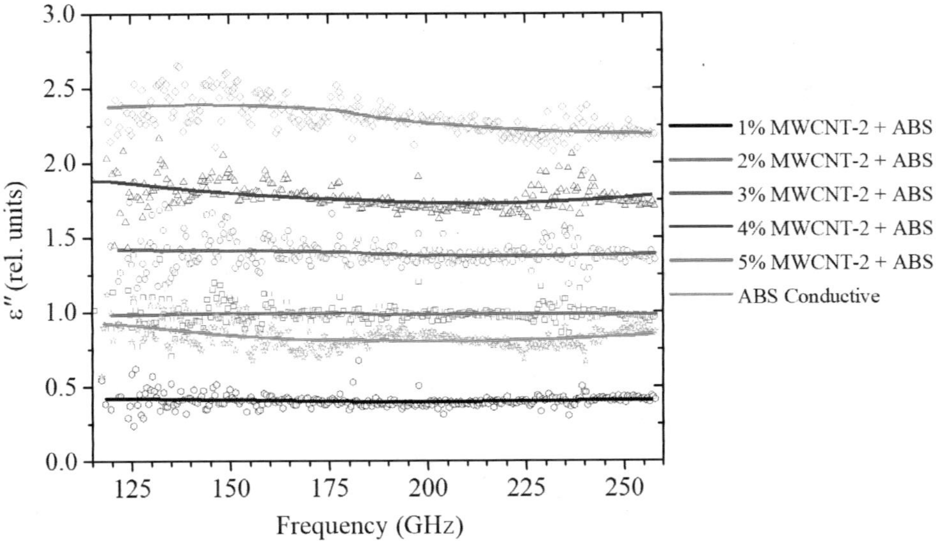

Figure 6. Frequency dependence of the imaginary part of permittivity.

4. Conclusions

New composite radiofilaments with multi-walled carbon nanotubes for 3D printing have been created. The measured dielectric constant of the samples showed a good agreement between the results in the SHF and EHF ranges. The results show that ABS Conductive plastic has the best absorbing properties on microwave frequency, and composite plastics with MWCNTs of 2% or more on EHF.

Acknowledgment

The research was supported by The Tomsk State University competitiveness improvement programme.

References

[1] Laur V, Kaissar Abboud M, Maalouf A, Palessonga D, Chevalier A andVille J 2019 *30th Asia-Pacific Microwave Conference Proceedings (APMC)* 1318–1320
[2] Pa P, Larimore Z, Parsons P and Mirotznik M 2015 Multi-material additive manufacturing of embedded low-profile antennas *Electronics Letters* **51 (20)** 1561–1562
[3] Wu Y, Isakov D and Grant P S 2017 Fabrication of Composite Filaments with High Dielectric Permittivity for Fused Deposition 3D Printing *Materials* **10** 1–12
[4] Badin A V, Kuleshov G E, Dorozhkin K V, Dunaevskii G E, Suslyaev V I and Zhuravlev V A 2018 *43rd International Conference on Infrared, Millimeter, and Terahertz Waves, IRMMW-THz* 8509938
[5] Grant P S, Castles F, Lei Q, Wang Y, Janurudin J M, Isakov D, Speller S, Dancer C and Grovenor C R M 2015 *Philosophical Transactions of the Royal Society A: Mathematical, Physical and Engineering Sciences* **373(2049)** 20140353
[6] Chalapat K, Sarvala K, Li J and Paraoanu G 2009 IEEE Transactions on **57** 2257–2267

IOP Conf. Series: Materials Science and Engineering **731** (2020) 012015 doi:10.1088/1757-899X/731/1/012015

Improving the Mechanical Properties of SiC-ceramics by means of Vacuum Electron-ion-plasma Alloying with Titanium

A A Leonov[1,2], Yu F Ivanov[1,2], M P Kalashnikov[2,3], E V Abdulmenova[2,3], V V Shugurov[1] and A D Teresov[1]

[1]Institute of High Current Electronics SB RAS, Tomsk, 634055, Russia
[2]National Research Tomsk Polytechnic University, Tomsk, 634050, Russia
[3]Institute of Strength Physics and Materials Science SB RAS, Tomsk, 634055, Russia

Email: laa91@tpu.ru

Abstract. The investigation results of elemental and phase composition, state of defective substructure and microhardness of the surface layer of «film (Ti)/substrate (SiC-ceramics)» system (Ti film 0.5 μm thick was deposited on the surface of SiC-ceramics) subjected to treatment with an intense pulsed low-energy electron beam (15 J/cm^2, 200 μs, 0.3 s^{-1}, 20 pulses) are presented. It is shown that irradiation of the «film (Ti)/substrate (SiC-ceramics)» system with an electron beam is accompanied by the formation of multielement multiphase (SiC; TiC; Ti_5Si_3) surface layer having submicro- and nanocrystalline structure. Microhardness of the irradiated surface layer reaches a value of 74 GPa, that is twice the value of microhardness of SiC-ceramics (36 GPa).

1. Introduction

Silicon carbide (SiC) ceramic is a substantial structural material applied in fields as aerospace, machinery and electronics due to its superior mechanical properties and chemical stability at high temperature, low density and coefficient of thermal expansion (CTE) [1-4]. In particular, this property mixture makes them promising candidates as lightweight alternative for several aircraft turbine components instead of super alloys. Recently, the application of first SiC parts was demonstrated in the hot section of new jet engines [5, 6]. However, the fabrication of complex SiC ceramic shaped components utilized in practical engineering was limited by its inherent nature of high brittleness, low malleability and poor machinability [7, 8]. Many strategies have been proposed to tackle this problem and adding reinforcing materials as a second phase in the form of particles [9-12], platelets/flakes [13-15], nanotubes [16-20], whiskers/fibers [21-25] is a useful method. But in most cases, in order to increase the mechanical properties of ceramics, it is only necessary to modify the surface layer [26-29]. The modifying method of the material surface with intense pulsed low-energy (to 30 keV) electron beams of submillisecond duration is the modern approach of the controlling microstructure and properties of parts and products. Impact of such beams on metals, alloys, cermet and ceramic materials is accompanied with the formation of an amorphous, nano and submicrocrystalline structure in the surface layer of the product, contributing to significant improvement the physicochemical, electrophysical, strength and many other properties [30-33].

The aim of work – revealing the regularities in the formation of the structure and properties of the surface layer of SiC-ceramics subjected to titanium alloying by irradiating the «Ti/SiC-ceramic» system with an intense pulsed electron beam.

2. Materials and methods

SiC powder with particle sizes (0.9 – 4.0) μm and nanopowder of the same composition (3 wt.%, particle size 60 nm) were used to fabrication ceramic samples. Samples with a diameter of 14 mm and thickness of 3 mm were obtained by SPS-sintering (SPS-515S (SPS SYNNEX), NR TPU, Tomsk). Sintering mode: sintering temperature 2100 °C, pressure up to 70 MPa and sintering time 10 min. The deposition of a 0.5 μm thick titanium film was carried out on a «KVINTA» unit (IHCE SB RAS) by vacuum electric arc spraying with plasma assistance («PINK» plasma generator) of a target of technically pure titanium of the VT1-0 grade. The «film (Ti)/substrate (SiC-ceramics)» system was irradiated with an intense pulsed electron beam at the «SOLO» device (IHCE SB RAS) with beam parameters: the accelerated electron energy is 18 keV, energy density of the electron beam is 15 J/cm^2, pulse duration 200 μs, quantity of pulses 20 at a repetition rate of 0.3 s^{-1}; the pressure of the residual gas (argon) in the working chamber is 10^{-2} Pa. Investigations of the structural-phase state of the modified surface layer of ceramics before and after the effect of the electron beam were carried out using the equipment of the Nano-Center TPU and the Center for Collective Use of the ISPMS SB RAS «NANOTECH» using scanning electron microscopy (JSM-7500FA, JEOL) and transmission diffraction electron microscopy (JEM-2100F, JEOL), X-ray diffractometer (XRD-7000S, Shimadzu). Microhardness of the modified layer was determined on the «PMT-3M» device.

3. Results and discussion

Vacuum electric arc spraying with plasma assistance («PINK» plasma generator) of a target of technically pure titanium of the VT1-0 grade is accompanied by the formation of a titanium film containing particles of a drop fraction on the surface of SiC ceramic (Figure 1a). The sizes of drops are changing from tens nanometers to units of micrometers. Irradiation of the «film (Ti)/substrate (SiC-ceramics)» system with an intense pulsed electron beam leads to melting of the titanium film (Figure 1b). On the surface of samples, regions with sizes up to 100 μm having dendritic crystallization structure are formed (Figure 2). Apparently, these regions were formed as a result of melting and high-speed crystallization of the droplet fraction particles.

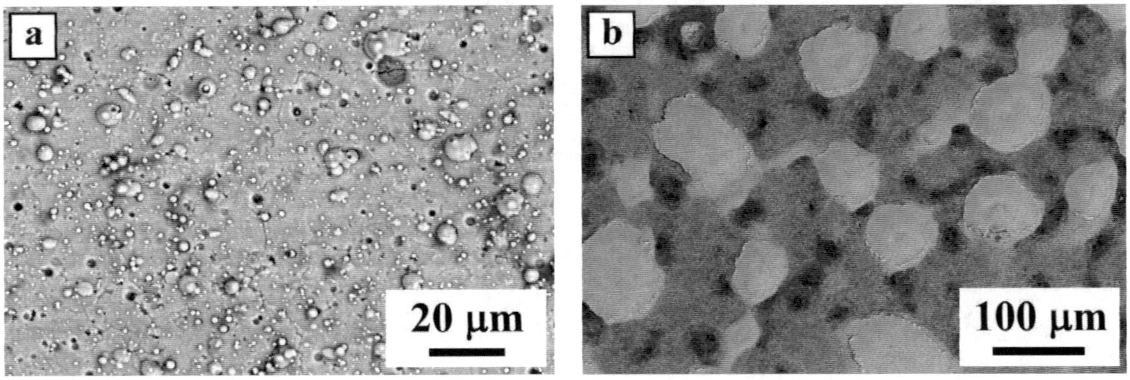

Figure 1. SEM images of the surface of SiC-ceramics samples with a deposited titanium film before (a) and after (b) irradiation with an intense pulsed electron beam.

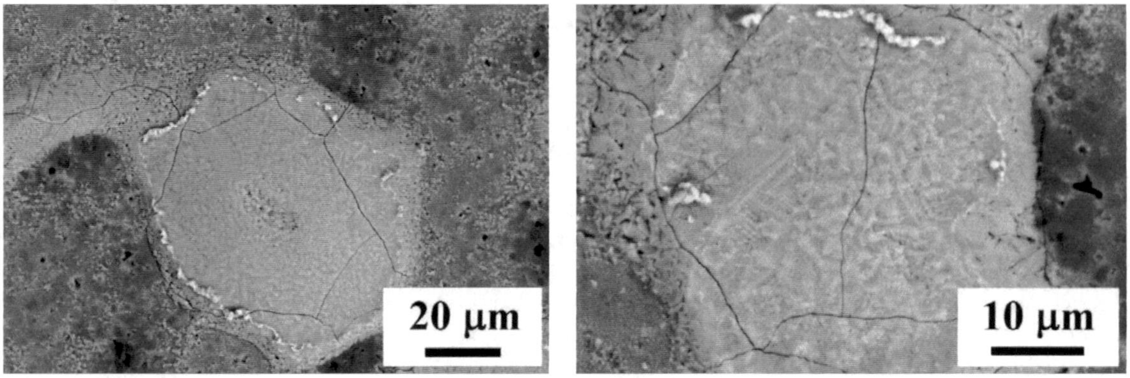

Figure 2. SEM images of surface of the «film (Ti)/substrate (SiC-ceramics)» system irradiated with an intense pulsed electron beam.

Concentration of elements in the surface layer of the «film (Ti)/substrate (SiC-ceramics)» system was determined by the energy-dispersive X-ray spectroscopy (EDS). It is established that the regions of dendritic crystallization (areas 1 and 3 in Figure 3a) are enriched in titanium atoms. Space separating these regions (areas 2, 4 and 5 in Figure 3a) is enriched with silicon and carbon and also contain titanium. On the average over the surface area of sample (Figure 3b), the concentration of carbon, silicon and titanium atoms is 30.4 at.%, 33.3 at.% and 36.0 at.%, respectively.

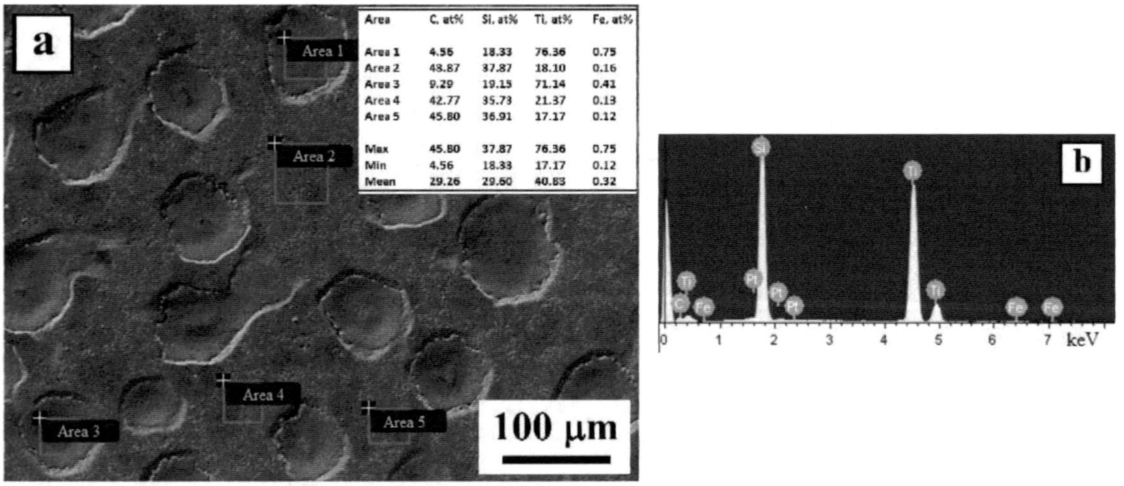

Area	C, at%	Si, at%	Ti, at%	Fe, at%
Area 1	4.56	18.33	76.36	0.75
Area 2	43.87	37.87	18.10	0.16
Area 3	9.29	19.15	71.14	0.41
Area 4	42.77	35.73	21.37	0.13
Area 5	45.80	36.91	17.17	0.12
Max	45.80	37.87	76.36	0.75
Min	4.56	18.33	17.17	0.12
Mean	29.26	29.60	40.83	0.32

Figure 3. Structure of the «film (Ti)/substrate (SiC-ceramics)» system irradiated with an intense pulsed electron beam (a). Areas of analysis of elemental composition are indicated. The table shows the elemental composition of these areas; (b) energy spectra obtained from the sample surface area given in (a).

The phase composition of surface layer of the «film (Ti)/substrate (SiC-ceramics)» system, irradiated with an intense pulsed electron beam was studied by X-ray diffraction analysis. As a result of performed studies, the presence of following phases was revealed: SiC – 18.5 %, TiC – 36.6 %, Ti_5Si_3 – 44.9 % (Figure 4). Therefore, electron beam treatment of the «film (Ti)/substrate (SiC-ceramics)» system is accompanied by melting of titanium film, doping the melt with silicon and carbon atoms, formation of new phases on the high-speed crystallization step and subsequent cooling.

Defective substructure of surface layer of the «film (Ti)/substrate (SiC-ceramics)» system irradiated with an intense pulsed electron beam was analyzed by transmission electron microscopy of thin foils. The foils were prepared by ionic thinning of plates cut perpendicular to the irradiation surface. This method of preparing foils (transverse foil method) allows sighting study the elemental and phase composition, defective substructure of material depending on the distance from the irradiation surface (Figure 5a).

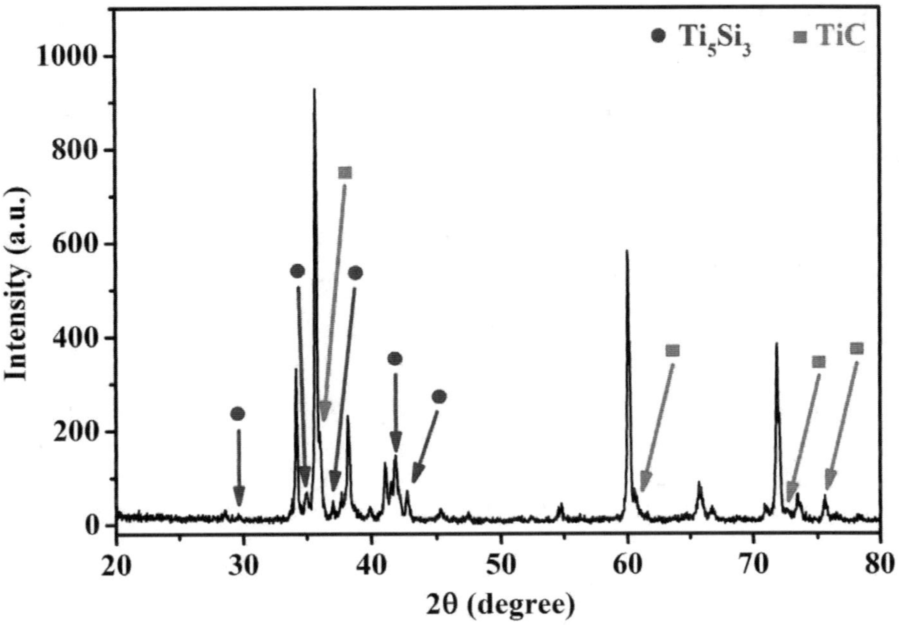

Figure 4. XRD pattern obtained from the surface layer of the «film (Ti)/substrate (SiC-ceramics)» system irradiated with an intense pulsed electron beam.

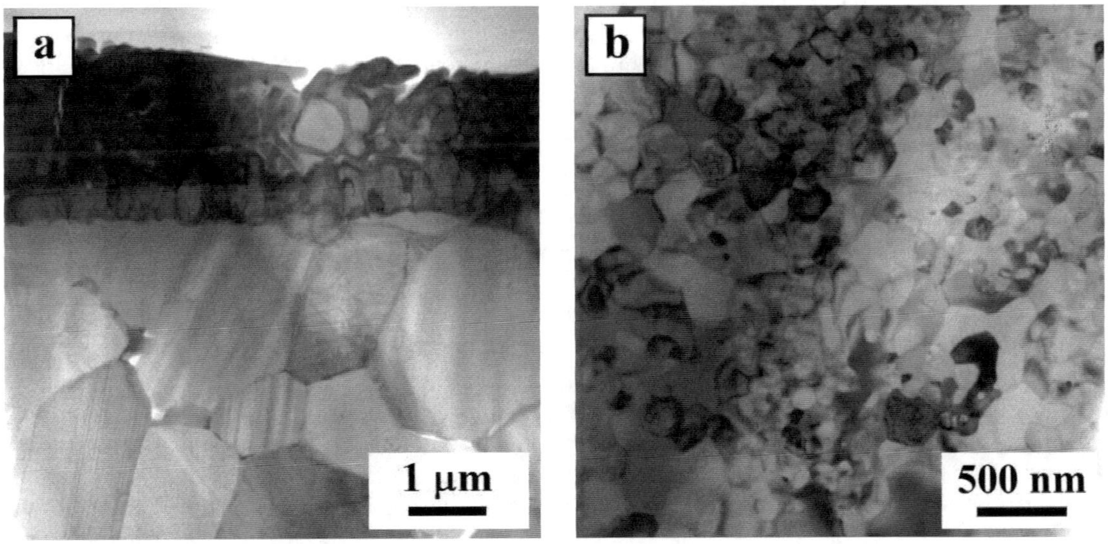

Figure 5. The regions of the structure formed during melting and high-speed crystallization, enriched with titanium atoms (regions containing the droplet fraction particle).

As a result of the investigations carried out, it was established that after irradiation with the electron beam of the «film (Ti)/substrate (SiC-ceramics)» system, surface layer is formed, the crystallite sizes of which range from 20 nm to 500 nm (Figure 5). Layer adjacent to the surface of SiC ceramic has a columnar structure (Figure 5a). Structure of the main volume of surface layer is formed by crystallites having a globular shape (Figure 5b).

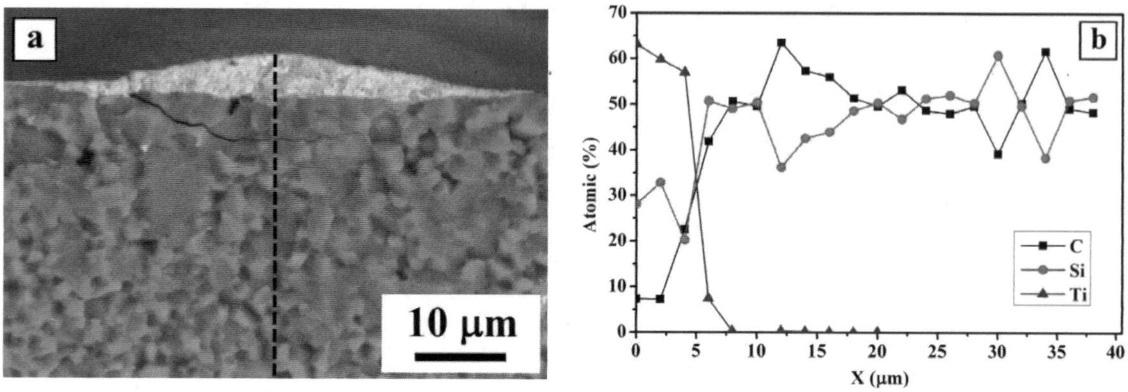

Figure 6. Structure of cross-section of the «film (Ti)/substrate (SiC-ceramics)» system (a); (b) the concentration profile of the main elements along the line indicated in (a).

Distribution of basic elements of ceramics modified layer was studied by the EDS analysis (Figure 6). It is established that high-speed melting of a titanium film leads to the formation of a multielement state: in the surface layer, there are of titanium, carbon and silicon atoms. Therein, titanium is the major element in the drop fraction particle (Figure 6b). When going across the «droplet/substrate» interface, the titanium concentration swiftly decreases. The thickness of titanium-doped layer of ceramic reaches 3 µm. Consequently, irradiation of the «film (Ti)/substrate (SiC-ceramics)» system with an intense electron beam is accompanied by the diffusion of ceramic elements into the deposited coating. Therefore, coating has a multiphase composition (titanium and silicon carbides, silicides and silicocarbides of titanium).

The formation of a nanocrystalline multielement multiphase surface layer as a result of irradiation of the «film (Ti)/substrate (SiC-ceramics)» system with a pulsed electron beam is accompanied by a significant increase of microhardness, the value of which reaches 74 GPa (microhardness of SiC ceramics is 36 GPa). The hardness maximum values are revealed near the boundary separating the regions enriched with titanium (droplets) from the main volume of the coating.

4. Conclusions

It is shown that irradiation of the «film (Ti)/substrate (SiC-ceramics)» system with an intense pulsed electron beam (15 J/cm^2, 200 µs, 0.3 s^{-1}, 20 pulses) leads to the formation of a surface layer, having a multielement multiphase structure of the submicro-nanocrystalline range. The mutual diffusion the atoms of film and substrate have been revealed. The thickness of titanium-doped layer of ceramic reaches 3 µm. It has been established that the microhardness of the «film (Ti)/substrate (SiC-ceramics)» system after irradiation with an electron beam reaches a value of 74 GPa, that is twice the value of microhardness of SiC-ceramics (36 GPa). The hardness maximum values are revealed near the boundary separating the regions enriched with titanium (droplets) from the main volume of the coating. It is obvious that such high values of microhardness are due to the formation of a nanocrystalline, multielement, multiphase state formed as a result of mutual diffusion of titanium, carbon and silicon atoms during irradiation with an intense pulsed electron beam.

Acknowledgment

This work was supported by the program of fundamental scientific research for 2019-2021. (Subject № 0291-2019-0002).

References

[1] Lee H-Y, et al. 2019 Structural and microstructural studies on SiC-SiO_2 ceramic composites *Materials Chemistry and Physics* **233** 203-212

[2] Feng D, et al. 2019 Pressureless sintering behaviour and mechanical properties of Fe_2O_3-

containing SiC ceramics *J. Alloy. Comp.* **790** 134-140
[3] King D S, et al. 2013 Silicon carbide–titanium diboride ceramic composites *J. Eur. Ceram. Soc.* **33** 2943-2951
[4] Grasso S, et al. 2015 Ultra-high temperature spark plasma sintering of α-SiC *Ceram. Int.* **41** 225-230
[5] Gardiner G 2015 *Aeroengine composites, part 1: the CMC invasion* (Cincinnati: Composite World, August, Gardner Business Media Inc) 38-41
[6] Raju K and Yoon D H 2016 Sintering additives for SiC based on the reactivity: a review *Ceram. Int.* **42** 17947-17962
[7] Kim Y W, et al. 2011 Electrodischarge–machinable silicon carbide ceramics sintered with yttrium nitrate *J. Am. Ceram. Soc.* **94** 991-993
[8] Eom J H, et al. 2015 Effect of additive composition on mechanical properties of pressureless sintered silicon carbide ceramics sintered with alumina, aluminum nitride and yttria *Met. Mater. Int.* **21** 525-530
[9] Buyakov A S, et al. 2018 Formation of pore structure in zirconia-alumina ceramics *Journal of Silicate Based and Composite Materials* **70** 27-31
[10] Nisar A, et al. 2019 Processing, microstructure and mechanical properties of HfB_2-ZrB_2-SiC composites: Effect of B_4C and carbon nanotube reinforcements *Int. J. Refr. Met. Hard Mat.* **81** 111-118
[11] Nayebi B, et al. 2019 Influence of vanadium content on the characteristics of spark plasma sintered ZrB_2–SiC–V composites *J. Alloy. Comp.* **805** 725-732
[12] Liao N, et al. 2020 Effects of nano ZrO_2 content on the comprehensive properties of BN-SiC composites *J. Alloy. Comp.* **813** 152180
[13] Zeller F, et al. 2017 Exceptional micromachining performance of silicon carbide ceramics by adding graphene nanoplatelets *J. Eur. Ceram. Soc.* **37** 3813-3821
[14] Buyakov A S, et al. 2019 Effects of low-modulus BN inclusions on properties of Y-TZP ceramic *Inorganic Materials: Applied Research* **10** 1159-1163
[15] Hanzel O, et al. 2019 Wire electrical discharge machinable SiC with GNPs and GO as the electrically conducting filler *J. Eur. Ceram. Soc.* **39** 2626-2633
[16] Leonov A A, et al. 2019 Keramicheskiy kompozit na osnove dioksida tsirkoniya, armirovannyy odnostennymi uglerodnymi nanotrubkami *Rossiyskiye nanotekhnologii* **14 (3-4)** 32-38 (in Russian) https://doi.org/10.21517/1992-7223-2019-3-4-32-38
[17] Popov O, et al. 2019 Reactive sintering of TiB_2-SiC-CNT ceramics *Ceram. Int.* **45** 22769-22774
[18] Shahedi Asl M, et al. 2016 Characteristics of multi-walled carbon nanotube toughened ZrB_2–SiC ceramic composite prepared by hot pressing *Ceram. Int.* **42** 1950-1958
[19] Leonov A A and Abdulmenova E V 2019 Alumina-based composites reinforced with single-walled carbon nanotubes *IOP Conf. Ser.: Mater. Sci. Eng.* **511** 012001
[20] Leonov A A, et al. 2017 Spark plasma sintering of ceramic matrix composite based on alumina, reinforced by carbon nanotubes *IOP Conf. Ser.: Mater. Sci. Eng.* **286** 012034
[21] Li S, et al. 2013 Fabrication and characterization of SiC whisker reinforced reaction bonded SiC composite *Ceram. Int.* **39** 449-455
[22] Song N, et al. 2017 Effects of SiC whiskers on the mechanical properties and microstructure of SiC ceramics by reactive sintering *Ceram. Int.* **43** 6786-6790
[23] Leonov A 2019 Effect of alumina nanofibers content on the microstructure and properties of ATZ composites fabricated by spark plasma sintering *Materials Today: Proceed.* **11** 66-71
[24] Cheng T, et al. 2019 Tensile properties of two-dimensional carbon fiber reinforced silicon carbide composites at temperatures up to 1800 C in air *Extr. Mech. Let.* **31** 100546
[25] Kim K A, et al. 2019 Composite ceramics based on silicon carbide with layered location of reinforcing SiC fibers *IOP Conf. Ser.: Mater. Sci. Eng.* **525** 012082
[26] Li B S, et al. 2014 Evolution of strain and mechanical properties upon annealing in He-implanted 6H-SiC *J. Nucl. Mater.* **455** 116-121

[27] Li J, et al. 2014 Evolution of amorphization and nanohardness in SiC under Xe ion irradiation *J. Nucl. Mater.* **454** 173-177

[28] Xu C L, et al. 2012 A HRXRD and nano-indentation study on Ne-implanted 6H–SiC *Nucl. Instr. Meth. Phys. Res. Sect. B* **286** 129-133

[29] Yang Y T, et al. 2018 Dose dependence of nano-hardness of 6H-SiC crystal under irradiation with inert gas ions *Nucl. Instr. Meth. Phys. Res. Sect. B* **422** 50-53

[30] Ivanov Y, et al. 2018 High chromium steel modification by the intense discrete electron beam: Structure and properties *Key Engineering Materials* **781** 64-69

[31] Leonov A A, et al. 2018 Structure and properties of the surface layer of «Ti/SiC-ceramic» system irradiated by low-energy pulsed electron beam *J. Phys.: Conf. Ser.* **1115** 032040

[32] Regula G and Yakimov E B 2016 Effect of low energy electron beam irradiation on Shockley partial dislocations bounding stacking faults introduced by plastic deformation in 4H-SiC in its brittle temperature range *Superlatt. Microstruct.* **99** 226-230

[33] Ivanov Yu, et al. 2018 Multilevel hierarchical structure formed in the film (Ti)/substrate (SiC-ceramics) system under irradiation by an intense pulsed electron beam *AIP Conference Proceedings* **2051** 020110

IOP Conf. Series: Materials Science and Engineering **731** (2020) 012016 doi:10.1088/1757-899X/731/1/012016

Electromagnetic Properties of Hexaferrite Polymer Composite Materials

A V Sbrodov

Department of Radioelectronics, Faculty of Radiophysics, National Research Tomsk State University, Tomsk, Russia

E-mail: sbrodov_av@niipp.ru

Abstract. In this paper the results of studies of the electromagnetic characteristics of composite radio materials based on Z-, W-, M-type barium ferrites and epoxy resin are presented. The frequency dependences of the reflection, transmission, and absorption coefficients are shown. The obtained results of the electromagnetic response are in good agreement with the published data on the frequencies of natural ferromagnetic resonance for the types of ferrites considered. Based on the obtained data, the spectra of complex dielectric and magnetic permeabilities were calculated. It has been suggested that the above changes in the complex magnetic permeability are associated with the presence of domain wall resonance in the region of 1 GHz and natural ferromagnetic resonance at higher frequencies.

1. Introduction

Modern radio equipment and electronics are widely used in everyday life. So, various wireless transceivers are widely used. These are Bluetooth, Wi-Fi routers and adapters, cell phones, smart home appliances and equipment (IoT-devices), these are all kinds of devices and systems used for geo-positioning and tracking of targets, radar stations, monitoring and control systems for robotic complexes. The frequency range of operation of such devices is very often in the range of microwave radiation [1-3]. But the use of this equipment creates a number of negative factors, such as a threat to biological objects, problems of electromagnetic compatibility and safety, as well as "pollution" of the general electromagnetic background. Composite materials that act as absorbers of microwave energy or shielding materials are widely used to solve this type of problem [4, 5]. Hexagonal ferrimagnets of various types are quite widely used as fillers for radio composites in the microwave range [6, 7]. Therefore, the study of the properties of such materials remains an urgent task.

2. Materials and Methods

2.1. Materials

The choice of matrix or binder composite is based on its purpose. Various types of sealants, varnishes, paints, epoxies, as well as rubbers of various types can act as a polymer matrix for samples of composite material. In our case, the base was an epoxy resin EDP-20 with a hardener made of polyethylene polyamine (PEPA). Barium ferrite powders were used as the active phase: $BaAlFe_{11}O_{19}$ (M type), $BaCo_{0.6}Zn_{1.4}Fe_{16}O_{27}$ (W type), $BaCo_{0.7}Zn_{1.3}Fe_{16}O_{27}$ (W type), $BaCo_{2.4}Ti_{0.4}Fe_{23.2}O_{41}$ (Z type).

IOP Conf. Series: Materials Science and Engineering **731** (2020) 012016 doi:10.1088/1757-899X/731/1/012016

Ferrites were obtained using SHS technology, followed by firing at 1200 °C and processing in a planetary mill. After grinding, the particle size was less than 100 microns.

2.2. *Obtaining experimental samples*

The manufacture of samples of composites was carried out according to the traditional technology of mixing the starting components. The sample production scheme is shown in Figure 1. After selecting the filler, we carefully weighed the components of the composite on a Shimadzu AUX-320 balance (error ~ 0.5 mg). This was followed by mixing the mixture to a homogeneous state with simultaneous sonication with a power of 20 W for 1 minute. Then, the resulting mixture was poured into special molds and polymerized at room temperature. After polymerization, the samples were machined with grinding machines to give the desired geometric shape. As a result, experimental samples were obtained in the form of washers with an inner diameter d_{in} = 3 mm, external diameter d_{ext} = 7 mm and thickness h = 2.25 mm.

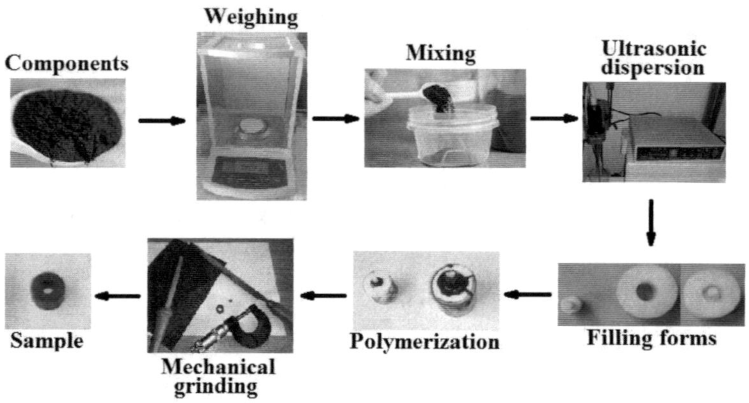

Figure 1. Production of samples of composite materials.

Thus, 4 experimental samples were prepared. The samples contained 40 wt.% epoxy resin and 60 wt.%. ferrite powders: $BaAlFe_{11}O_{19}$, $BaCo_{0.6}Zn_{1.4}Fe_{16}O_{27}$, $BaCo_{0.7}Zn_{1.3}Fe_{16}O_{27}$, $BaCo_{2.4}Ti_{0.4}Fe_{23.2}O_{41}$.

2.3. *Measuring equipment*

The electromagnetic properties of the obtained composites were measured by the waveguide method with the inclusion of a coaxial cell with an inner diameter $d_{int.}$ = 3.05 mm and external diameter $d_{ext.}$ = 7 mm. The vector network analyzer P4M-18 produced by the company "Micran" acted as a measuring installation. The measurements were carried out according to the "for pass" scheme (Figure 2). As a result, were obtained the family of frequency dependences of the reflection coefficients (R), transmission (T), and absorption (A) in the frequency range from 10 MHz to 18 GHz.

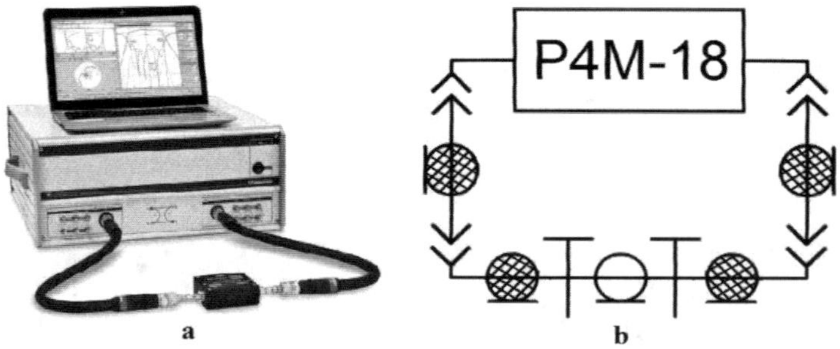

Figure 2. Measuring installation: a) vector network analyzer P4M-18; b) Schematic diagrams of measurements "on pass".

The "on pass" (Figure 2b) measurement scheme allows you to measure the values of the coefficients of the scattering matrix. In other words, to determine the value of parameters *S*11, *S*12, *S*21, *S*22. These S-parameters determine the coefficients of reflection and transmission of an electromagnetic wave through a sample placed in the measuring cell, with forward direction and reverse passage. This device (Figure 2a) allows you to track the phase change of the obtained coefficients of the scattering matrix. Based on the measured *S*-parameters, as well as their phase, it is possible to calculate the spectra of complex magnetic and dielectric constants.

3. Results and discussion

Figure 3 shows the frequency dependences of the reflection coefficients (*R*), transmission (*T*), and absorption (*A*) for the studied composite samples.

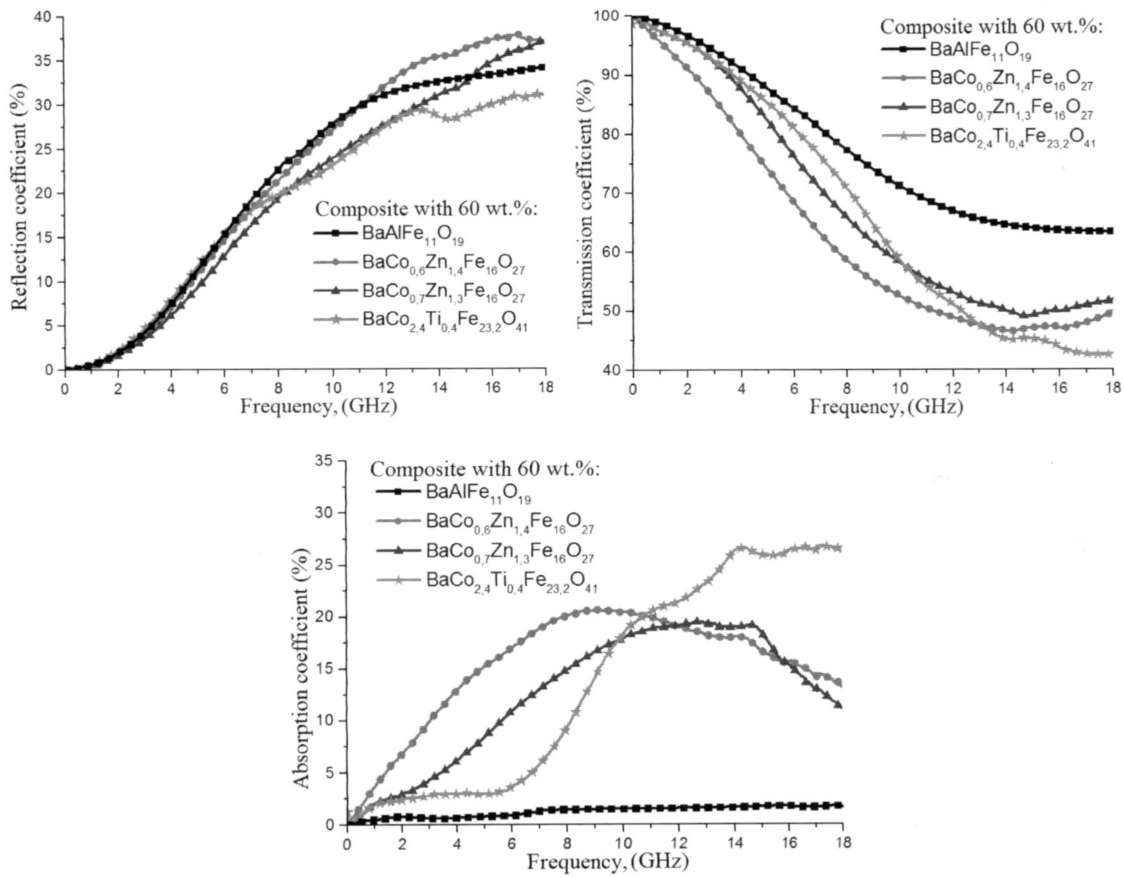

Figure 3. Frequency dependences of the reflection coefficients (*R*), transmission (*T*), absorption (*A*) for composite materials based on epoxy resin and ferrites.

Analyzing the obtained dependences in Figure 3, it can be noted that with an increase in the frequency for all samples, a decrease in the transmission coefficient is noticeable. The sample containing $BaAlFe_{11}O_{19}$ powder has the lowest shielding level (up to 65%). The samples with $BaCo_{0.6}Zn_{1.4}Fe_{16}O_{27}$ and $BaCo_{2.4}Ti_{0.4}Fe_{23.2}O_{41}$ have the maximum of shielding level (up to 45%). This is due to an increase in the reflection and absorption coefficients. Thus, with increasing frequency, an almost linear increase in the reflection coefficient is observed. Moreover, for all ferrites, its value is approximately the same, this is due to the same thickness of the samples and close electromagnetic parameters. The highest absorption at frequencies up to 10 GHz is possessed by a sample containing $BaCo_{0.6}Zn_{1.4}Fe_{16}$, and above 10 GHz a sample with $BaCo_{2.4}Ti_{0.4}Fe_{23.2}O_{41}$, whose absorption coefficient for this reaches 27%. The lowest absorption coefficient has a sample based on $BaAlFe_{11}O_{19}$ and reaches 2-3%. This behavior of the electromagnetic response is in good agreement with published data on the frequencies of natural ferromagnetic resonance (NFMR) for the considered types of ferrites.

Based on the measurements of the *S*-parameters, the complex permittivity and permeability were calculated. The calculation of electromagnetic parameters was carried out according to the Becker-Jarvis method using a program written in the Mathcad environment, based on the content of the article "Wideband Reference-Plane Invariant Method for Measuring Electromagnetic Parameters of Materials" [8]. So, Figure 4 shows the frequency dependences of the complex dielectric and magnetic permeabilities for samples of composites based on epoxy resin and ferrimagnets.

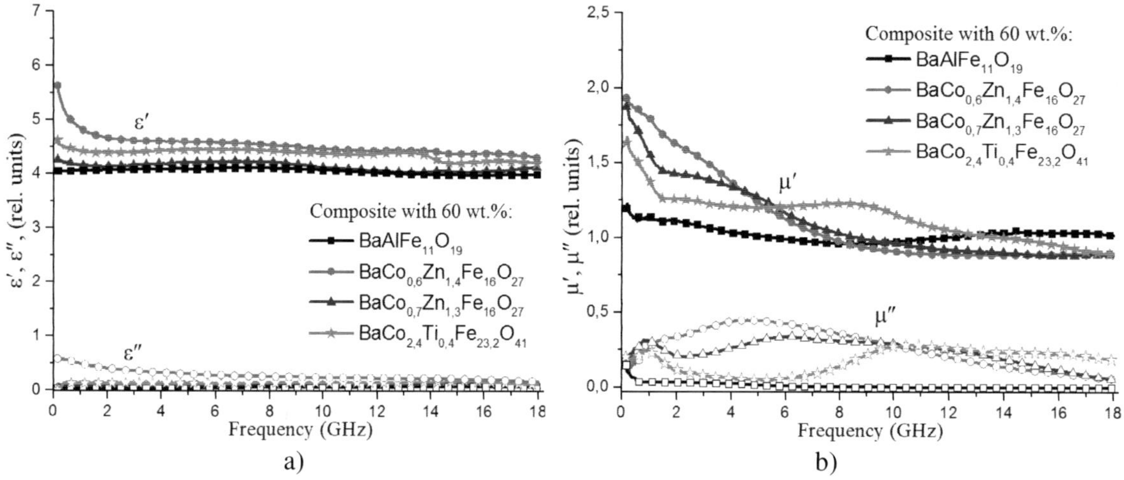

Figure 4. Frequency dependences of permeability (a) and permittivity (b) for a composite material based on epoxy resin and ferrites.

Analyzing the spectra of complex permittivity and magnetic permeability, it can be noted that at low frequencies, the highest ε' and ε" values belong to $BaCo_{0.6}Zn_{1.4}Fe_{16}O_{27}$, reaching ε' = 5.6 rel. units and ε" = 0.6 rel. units. At frequencies above 4 GHz all the materials under consideration have a similar dependence and the value of ε' belongs in the range from 4 rel. units up to 5 rel. units, and ε" up to 0.5 rel. units. On the frequency dependences of the complex magnetic permeability, there is a decrease in the real part and maxima on the imaginary part. Such changes in the complex magnetic permeability are probably associated with the presence of resonance of domain walls in the region of 1 GHz and natural ferromagnetic resonance at higher frequencies.

4. Conclusions

From the obtained spectra of the complex dielectric and magnetic permeabilities. W-type hexaferrites are more preferred to use in the microwave range at frequencies up to 5 GHz, because the natural ferromagnetic resonance of these materials is in this frequency range. From 6 to 12 GHz, Z-type hexaferrites demonstrate the better properties than W-type hexaferrites. M-type hexaferrite works at higher frequencies, which follows from the literature on its natural ferromagnetic resonance, which should be above 50 GHz for this structure.

Acknowledgment
The author is grateful to PhD V.A. Zhuravlev of the Department of Radioelectronics of National Research Tomsk State University (Tomsk) for help in the calculation of electromagnetic parameters. The study was sponsored by RFBR in the framework of a research project №18-32-00810.

References
[1] Shakhnovich I 2015 5G wireless systems: the telecommunication paradigm that will change the world. Brief Abstracts *Electronics: Scientific and Technical Library* **7** 48–55
[2] Daniel S,Green D, Dohrman C, Chang T 2015 Compound Semiconductor Technology for Modern RF Modules: Status an Future Directions *CS MANTECH Conference 2015* **21** 11-14
[3] Maltsev P, Shakhnovich I 2015 Microwave technology - the foundation of the electronics of the future. Trends and Markets *Electronics: Scientific and Technical Library* **8** 72–84
[4] Kuleshov G E, Zhuravleva Y V, Chernobrova D A 2014 Microwave electromagnetic arameters of composite radiomaterials based on carbon nanostructures and ferrimagnetics *International Conference of Young Specialists on Micro/Nanotechnologies and Electron Devices, EDM* 158-160
[5] Suslyaev V, Kazmina O, Kuleshov G, Korovin E, Dorozhkin K, Lebedeva E 2017 Electromagnetic response of the three-layer construction on the basis of barium hexaferrite and a foam glass *IOP Conference Series: Materials Science and Engineering* **168** 012099
[6] Badin A V, Dorozhkin K V, Kuleshov G E, Zhuravlev, V A, Suslyaev V I, Dunaevskii G E, Bilinskii K V 2018 Ferromagnetic resonance in hexagonal ferrite $BaFe_{12}O_{19}$ at the EHF frequency range *International Conference on Infrared, Millimeter, and Terahertz Waves, IRMMW-THz* 8509938
[7] Nikmanesh H, Hoghoghifard, Hadi-Sichani B 2019 Study of the structural, magnetic, and microwave absorption properties of the simultaneous substitution of several cations in the barium hexaferrite structure *Journal of Alloys and Compounds* **775** 1101–1108
[8] Chalapat K, Sarvala K, Li J, Sorin G 2009 Wideband Reference-Plane Invariant Method for Measuring Electromagnetic Parameters of Materials *IEEE Transactions on Microwave Theory and Techniques* **57** 2257–2267

IOP Conf. Series: Materials Science and Engineering **731** (2020) 012017 doi:10.1088/1757-899X/731/1/012017

Influence of Heterogeneity Composite Filament for Additive Technology on its Electrophysical Properties at THz Range

P P Smygalina, A V Badin, D A Pidotova and E A Trofimov

Faculty of Radiophysics, Tomsk State University, 36, Lenina av., Tomsk, 634050, Russia

thzlab@mail.ru

Abstract. In paper the results of a research of the effect of porosity on the dielectric properties of a composite filament for a 3D-printer based on the fused deposition method were presented. The process of creating a composite filaments based on acrylonitrile butadiene styrene and multi-walled carbon nanotubes is described. On a quasi-optical spectrometer, the dielectric permittivity of samples of composite materials 3D printed from a porous filament at a frequency of 210 GHz were obtained.

1. Introduction

Nowadays there is growing interest in using the THz frequency range. Difficulties in the practical development of the THz range are associated with insufficient development of the hardware element base. Microwave technology loses efficiency and reduces the functionality of measuring instruments. The development and production of THz structures today is difficult due to limitations in the technology for their production. The development of alternative devices is based on the use of new materials that effectively interact with electromagnetic radiation in the THz range [1]. As such materials graphene [2-5], single-walled [6] and multi-walled nanotubes (MWCNT) [7-14] are used for the construction of sensors [15-20] can be applied. In that regard the task to study the properties of materials that might be suitable for using at this frequency range. In this work, in order to create a sample suitable for THz research, a 3D printer based on the method of fused deposition is used. One of the ways to get filaments for printing is screwless hot extrusion technology. This technology is critical to the feedstock, because heterogeneities in the feedstock affects the final quality of the filament. One of the problem in the manufacture of filament is the porosity of the structure. It affects the electrophysical properties of the material. In this way it is actual to study the influence of porosity and density on the dielectric properties of the filament at THz range.

2. Experimental work

In this research, a composite filament based on acrylonitrile butadiene styrene (ABS) and 2% MWCNT with an average diameter of 18.6 nm was used as a material for 3D printing. For the research filaments with different numbers of extrusion iterations were made. The procedure for manufacturing the material for research is presented in Figure 1.

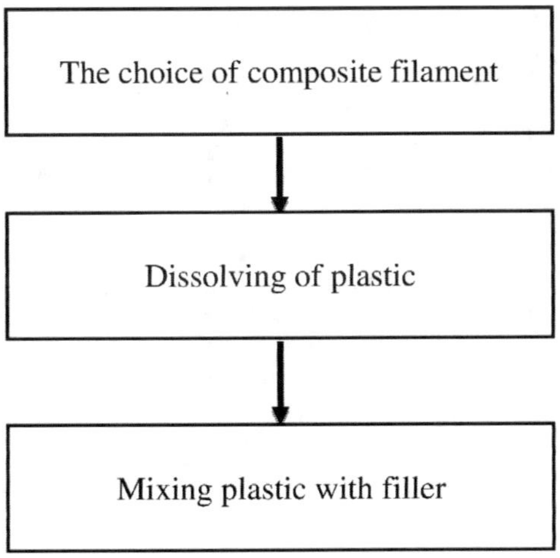

Figure 1. ABS/MWCNTs filament manufacturing process.

The research process is illustrated in Figure 2.

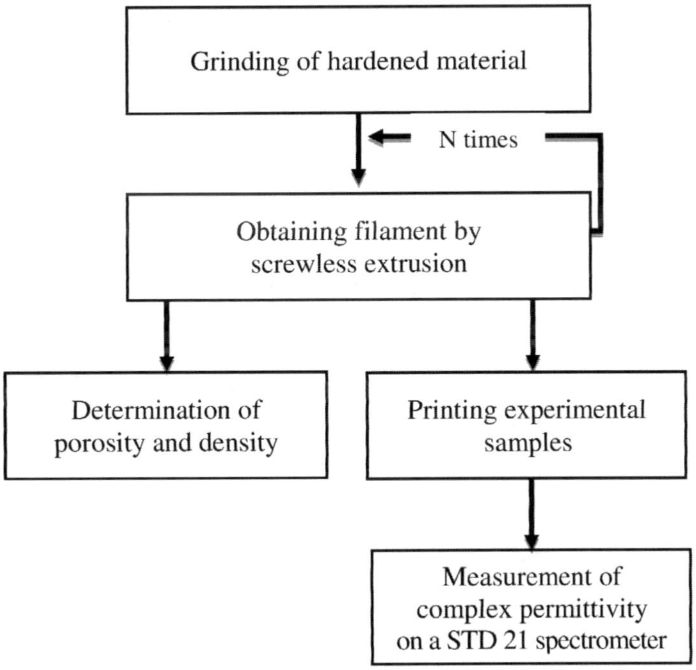

Figure 2. Block-diagram of research process.

The first stage of the study was to determine the correlation between density and porosity. Density measurements for six test samples of each filament were carried out. To determine the porosity of the material, a section of the filament was observed under a microscope. Then the obtained microphotographs in the photo editor (Figure 3) determined the percentage of pores to the total cut-off area. Six cut areas were examined for each sample. Table 1 presents the measured densities and porosities (physical quantities) of the obtained filaments.

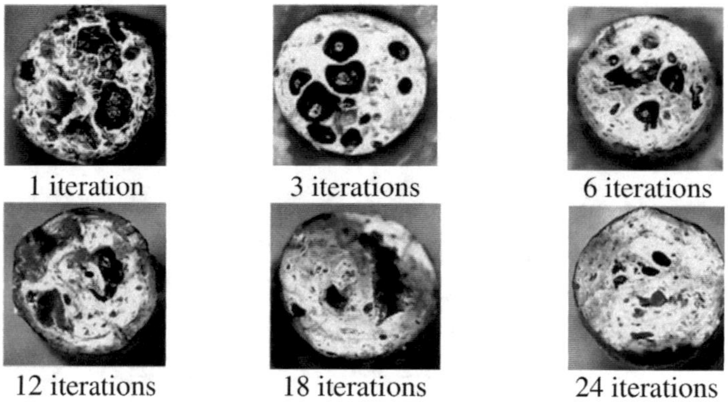

Figure 3. Microphotographs of sections of ABS/MWCNTs filaments of different orders of extrusion iteration.

With an increase in the number of iterations from 1 to 24, there is an increase in density from 0.7039 to 0.9956 g/cm^3 and a decrease in porosity from 39.33% to 14.17% (Figure 4).

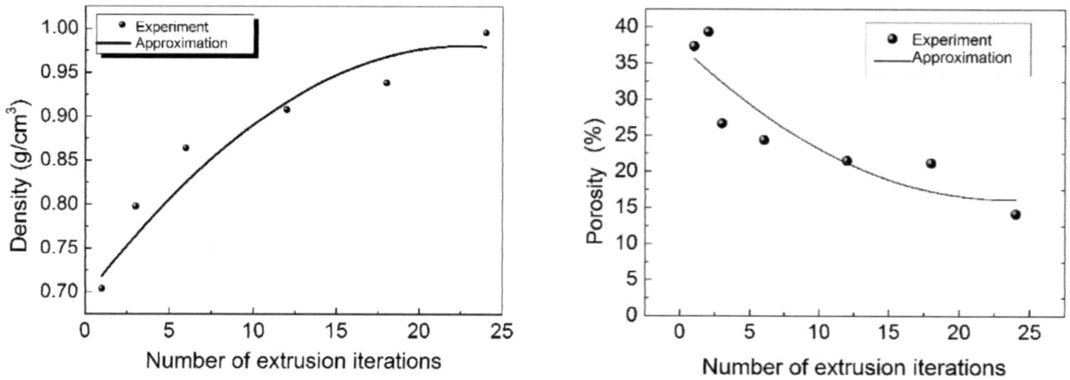

Figure 4. Dependence of the density (left) and porosity (right) of the ABS/MWCNTs composite material on the number of extrusion iterations.

The graph of porosity versus density shows a direct dependence of these values (Figure 5).

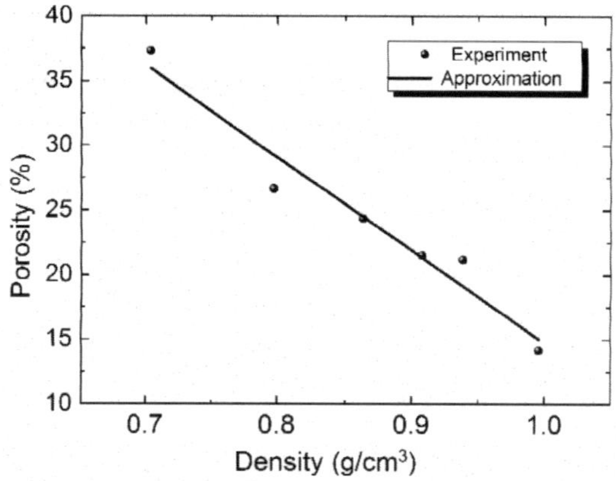

Figure 5. Dependence of material porosity on density of ABS/MWCNTs filaments.

IOP Conf. Series: Materials Science and Engineering **731** (2020) 012017 doi:10.1088/1757-899X/731/1/012017

The next step was printing of round plate samples with diameter 20 mm and thickness from 0.37 mm to 1.17 mm. On quasioptical spectrometer STD-21 at the frequency range of 115-258 GHz research of the dielectric properties of materials was carried out (Figure 6). A backward wave oscillator (BWO) was used as a source of electromagnetic radiation. THz radiation was detected by an acousto-optic converter (Golay cell).

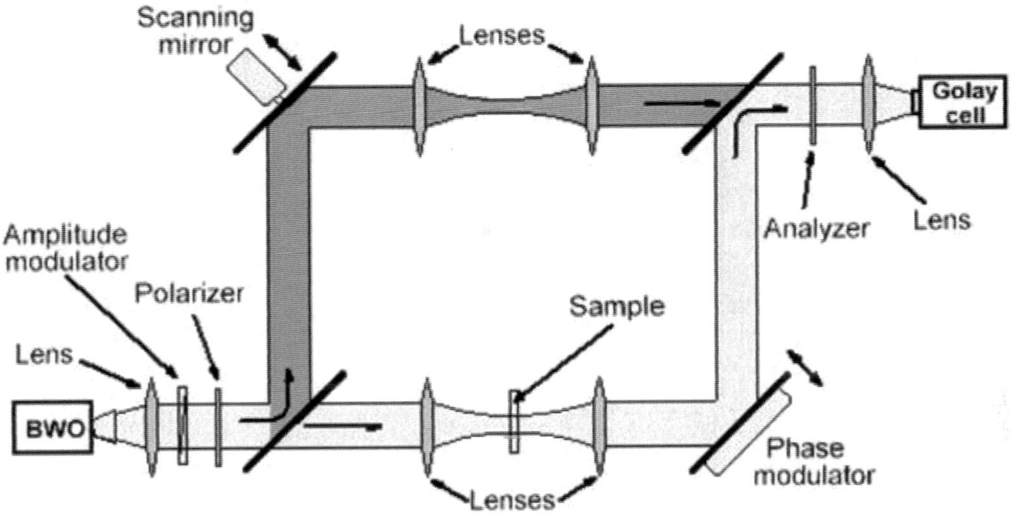

Figure 6. The experimental setup for research of electromagnetic properties of composite materials in THz frequency range.

In quasioptical spectrometer values of transmission coefficient $T(f)$ and phase shift $\varphi(f)$ were measured. Then the complex dielectric permeability was calculated.

$$T(f) = E\frac{(1-r)^2 + 4r\sin^2\beta}{(1-rE^2 + 4rE\sin^2(N+\beta)} \quad (1)$$

$$\varphi(f) = N + \mathrm{arctg}\frac{Er\sin(2(N+\beta))}{1-Er\cos(2(N+\beta))} + \mathrm{arctg}\frac{b}{a^2+b^2+a} - \mathrm{arctg}\frac{b}{a+1} \quad (2)$$

$$r = ((a-1)^2 + b^2)/((a+1)^2 + b^2) \quad (3)$$

$$N = 2\pi ndf / c \quad (4)$$

$$\beta = \mathrm{arctg}\left(2b/(a+b^2-1)\right) \quad (5)$$

The multiplier E takes into account the attenuation of the electromagnetic wave with physical parameters of the plane-parallel layer of material. The r-factor determines the reflection of the wave at the boundary of the air and material layer. N and β expressions take into account the phase shift, d is the thickness of the test sample, f is the frequency of electromagnetic radiation, c is the speed of light in vacuum. The dielectric permittivity values given in Table 1 are determined at a frequency of 210 GHz.

Table 1. Correlation of the number of extrusion iterations, average density, porosity and complex dielectric permeability of ABS/MWCNTs filament.

Number of extrusion iterations	Average density (g/cm^3)	Average porosity (%)	ε' (rel. units)	ε'' (rel. units)
1	0.7039	37.33	—	—
3	0.7976	26.67	2.36	1.30
6	0.8640	24.33	4.31	2.12
12	0.9080	21.50	4.87	2.67
18	0.9385	21.17	5.12	2.77
24	0.9956	14.17	4.82	2.13

Based on these values, was constructed the dependences of the real and imaginary parts of the dielectric permeability on the number of iterations of extrusion (Figure 7). Filaments subjected to a single iteration of extrusion were not suitable for printing, so no data were provided.

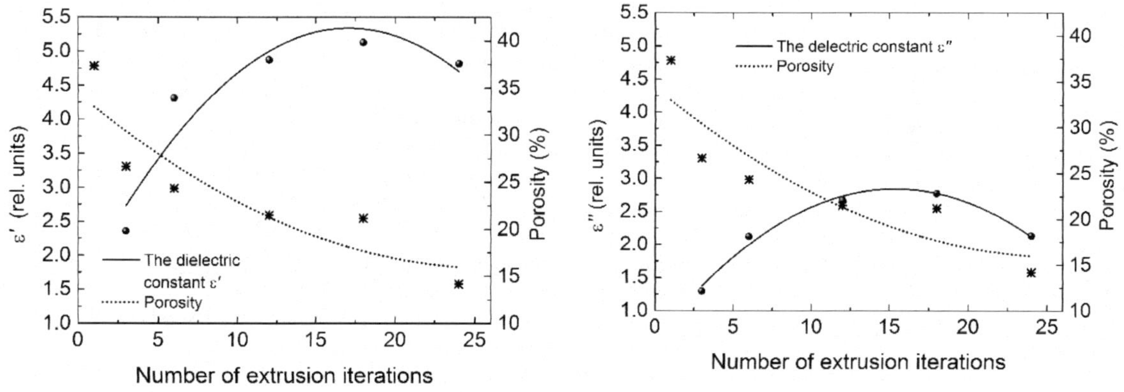

Figure 7. Dependence of real (left) and imaginary (right) parts of the dielectric constant and porosity on the number of extrusion iterations of ABS/MWCNT composite filament.

It was found that with increasing iterations of extrusion, the density and dielectric permeability of the composite filament increases and the porosity decreases. The increase in dielectric constant is observed due to the reduction of air volume in the pores. It is also worth noting that the dielectric permeability and porosity have a nonlinear inverse dependence (Figure 4).

In such a way, it is necessary to perform at least 12 iterations of extrusion in order to obtain composite filament with stable dielectric permeability values. In order to create a dense enough structure, it is necessary to perform only 18 iterations.

References

[1] Badin A V, Berdyugin A I, Vigovskii V Y and Dunaevskii G E 2017 System of automated measurement of electromagnetic response of anisotropic materials in quasi-optical beams *18th International Conference of Young Specialists on Micro/Nanotechnologies and Electron Devices, EDM* 301-4

[2] De la Osa G, Pérez-Coll D, Miranzo P, Osendi M I and Belmonte M 2016 Printing of Graphene Nanoplatelets into Highly Electrically Conductive Three-Dimensional Porous Macrostructures *Chem. Mater.* acs.chemmater.6b02662

[3] García-T͡On E, Barg S, Franco J, Bell R, Eslava S, D'Elia E, Maher R C, Guitian F and Saiz E 2015 Printing in three dimensions with Graphene *Adv. Mater.* **27** 1688–93

[4] Botta L, Scaffaro R, Sutera F and Mistretta M C 2018 Reprocessing of PLA/graphene nanoplatelets nanocomposites *Polymers (Basel)* **10**

IOP Conf. Series: Materials Science and Engineering **731** (2020) 012017 doi:10.1088/1757-899X/731/1/012017

[5] Dul S et al 2016 Graphene-ABS nanocomposites for fused deposition modelling *17th Eur. Conf. Compos. Mater. ECCM 2016.* № 7

[6] Gonzalez G, Chiappone A, Roppolo I, Fantino E, Bertana V, Perrucci F, Scaltrito L, Pirri F and Sangermano M 2017 Development of 3D printable formulations containing CNT with enhanced electrical properties *Polymer (Guildf)* **109** 246–53

[7] Sezer H K and Eren O 2019 FDM 3D printing of MWCNT reinforced ABS nanocomposite parts with enhanced mechanical and electrical properties *J. Manuf. Process.* **37** 339–47

[8] Szczurek A, Ortona A, Ferrari L, Rezaei E, Medjahdi G, Fierro V, Bychanok D, Kuzhir P and Celzard A 2015 Carbon periodic cellular architectures *Carbon N. Y.* **88** 70–85

[9] Kuleshov G E and Badin A V 2019 Study of electromagnetic characteristics of polymer materials based on single-walled and multi-walled carbon nanotubes *IOP Conf. Ser. Mater. Sci. Eng.* **525** 012030

[10] Badin A V., Kuleshov G E, Dorozhkin K V., Dunaevskii G E, Suslyaev V I and Zhuravlev V A 2018 Anisotropy of electrical properties of 3D-printing MWCNT composites at the THz frequency range *Int. Conf. Infrared, Millimeter, Terahertz Waves, IRMMW-THz* 5–6

[11] Postiglione G, Natale G, Griffini G, Levi M and Turri S 2015 Conductive 3D microstructures by direct 3D printing of polymer/carbon nanotube nanocomposites via liquid deposition modeling *Compos. Part A Appl. Sci. Manuf.* **76** 110–4

[12] Wu Y, Isakov D and Grant P S 2017 Fabrication of composite filaments with high dielectric permittivity for fused deposition 3D printing *Materials (Basel)* **10**

[13] Smygalina P P, Trofimov E A, Pidotova D A, Badin A V 2019 Assessment of the effect of porosity of a composite ABS filament with MWCNTs on its electrophysical properties in the EHF range *Proceedings of the Sixteenth All-Russian Conference of Student Research Incubators* 69-72

[14] Zhang J, Yang B, Fu F, You F, Dong X and Dai M 2017 Resistivity and Its Anisotropy Characterization of 3D-Printed Acrylonitrile Butadiene Styrene Copolymer (ABS)/Carbon Black (CB) Composites *Appl. Sci.* **7** 20

[15] Czyzewski J, Burzyński P, Gaweł K and Meisner J 2009 Rapid prototyping of electrically conductive components using 3D printing technology *J. Mater. Process. Technol.* **209** 5281–5

[16] Kwok S W, Goh K H H, Tan Z D, Tan S T M, Tjiu W W, Soh J Y, Ng Z J G, Chan Y Z, Hui H K and Goh K E J 2017 Electrically conductive filament for 3D-printed circuits and sensors *Appl. Mater. Today* **9** 167–75

[17] Flowers P F, Reyes C, Ye S, Kim M J and Wiley B J 2017 3D printing electronic components and circuits with conductive thermoplastic filament *Addit. Manuf.* **18** 156–63

[18] Guo S, Yang X, Heuzey M-C and Therriault D 2015 3D printing of a multifunctional nanocomposite helical liquid sensor *Nanoscale* **7** 6451–6

[19] Yoon J Y and Kim G W 2019 Harnessing the bilinear nonlinearity of a 3D printed biomimetic diaphragm for acoustic sensor applications *Mech. Syst. Signal Process.* **116** 710–24

[20] Hamzah H H, Shafiee S A, Abdalla A and Patel B A 2018 3D printable conductive materials for the fabrication of electrochemical sensors: A mini review *Electrochem. commun.* **96** 27–31

IOP Conf. Series: Materials Science and Engineering **731** (2020) 012018 doi:10.1088/1757-899X/731/1/012018

Quasi-Optical Resonant Single-Frequency Flaw Detector of Polymer Filament for 3D Printing

D D Teterina, I O Dorofeev and A V Badin

Faculty of Radiophysics, Tomsk State University, 36, Lenina av., Tomsk, 634050, Russia

teterinadar@gmail.com

Abstract. In the paper quasi-optical resonant single-frequency flaw detector for investigation of heterogeneity of 3D printer polymer filament are described. A scheme of quasi-optical flaw detector based on 10 GHz open resonator with a filament broaching mechanism is given. Industrial filaments based on polystyrene, polyethylene terephthalate glycol and acrylonitrile styrene acrylate were tested in a flaw detector. The measured amplitudes at the maximum of the resonance curve in the presence of a filament in the inside of the resonator are presented.

1. Introduction

Additive printing technologies are currently widely used in the creation of various objects. As for the applications of 3D printing in electronic equipment [1-4], in addition to the mechanical properties of such objects, the electrophysical properties [5-7] of the material used also play an important role. In this case, a joint solution of the problems is possible, both of creating load-bearing structures, and of electromagnetic compatibility and protection from radio emission of individual units and devices.

This approach requires the creation of an initial filament for 3D printing with specified electrophysical parameters [8-14]. It is also required to ensure the constancy of these parameters during the printing process. The best option here is to use continuous non-destructive testing of fiber throughout its length. It is essential that the control should be radio wave and be carried out at frequencies close to the operating frequency of the device in which the material will be applied. Given the small transverse dimensions of the fiber compared to the wavelength in the microwave ranges, it is advisable to use resonator methods. Sensors based on open resonators possess a number of important advantages when creating resonant sensors, which ensure the free placement of a controlled fiber section in a resonant volume, a sparse frequency spectrum, and high quality factor of oscillations in a wide frequency range. The free placement of an object is important not only in terms of ease of access to it, but also in terms of the optimal selection of its position in order to minimize the effects of its vibrations and movements in the resonant volume on the output signal.

2. Problem definition

In the simplest case, the filament diagnostic device consists of a highly stable microwave generator, open resonator [15-18], microwave oscillation detector, processing and indication device. The principle of its operation is to fix the change in the signal at the resonator output, which is due to a change in the filament parameters. Such device can operate in the measurement mode of parameter of the fiber, for example, the diameter with a constant value of the dielectric constant, or vice versa. In this case, calibration of the scale by reference samples is required. The second mode is fixing a small local (less than the diameter of the beam) heterogeneity of either diameter or permittivity. One of the main types of resonator oscillations is usually used, the deviation of the amplitude-frequency characteristic of which (the resonance curve) from the amplitude-frequency characteristic of the empty resonator is determined by the electrophysical characteristics of the introduced filament and their changes. Thus, the output signal level of a single-frequency converter is determined by tuning of the resonator and the properties of the portion of the fiber located in the resonant beam. To achieve maximum conversion steepness, various settings for fiber samples with different parameters are required.

3. Theory

3.1. Characteristics of a quasi-optical resonator with a polymer filament.

If the losses in the fiber are small, then the resonance curve of the resonator with the object relative to the empty cavity mainly undergoes a shift along the frequency axis (Figure 1a), since the diffraction loss at such an object is also small. In this case, it is advisable to tune the resonator so that the frequency of the generator is on the slope of the resonance curve of the resonator with the sample at about half power. This ensures maximum measurement of the output signal when the parameters of the stretched fiber deviate. The choice of a slope (low-frequency or high-frequency) is practically irrelevant here.

If the losses in the fiber lead to a noticeable broadening of the resonance curve relative to the resonance curve of the empty resonator, but its displacement along the frequency axis is small, it is advisable to combine the generator frequency with the peak of the resonance curve of the open resonator. The advantage here will be provided by a higher value of the signal itself.

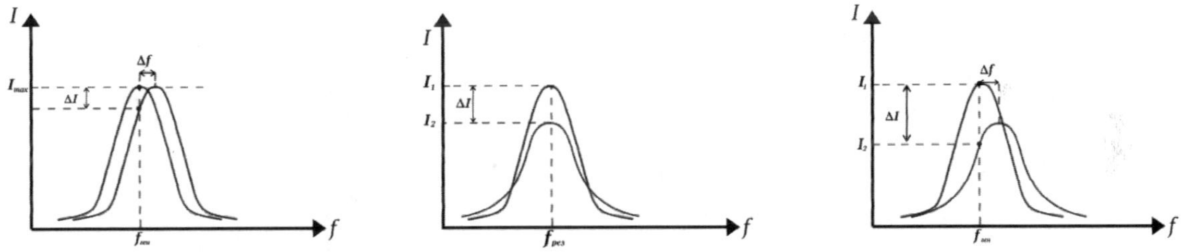

a) b) c)

Figure 1. Changes in the amplitude-frequency characteristic of an open resonator in the case of: a) a shift only along the frequency axis; b) only a change in amplitude; c) shear and broadening of the resonance curve.

Finally, if a change in the fiber parameters leads both to a shift of the resonance curve along the frequency axis and to its broadening (Figure 1c), then the point of optimal tuning may also be on the slope, but it is important to choose not only its position, but also the necessary slope (low-frequency, or high frequency).

3.2. System of quasi-optical resonant single-frequency flaw detector for heterogeneity of polymer filaments.

The device (Figure 2) contains a voltage-controlled oscillator (VCO) at a frequency of 10 GHz as a signal source, synchronized using a phase-locked loop in a reference highly stable crystal oscillator frequency (COF). To simplify the implementation of the circuit measuring the transmission coefficient of the measuring open resonator not only at the resonant frequency, but also at any given point in the resonance curve, it is advisable to use a highly stable non-tunable generator, and to perform tuning by the resonator.

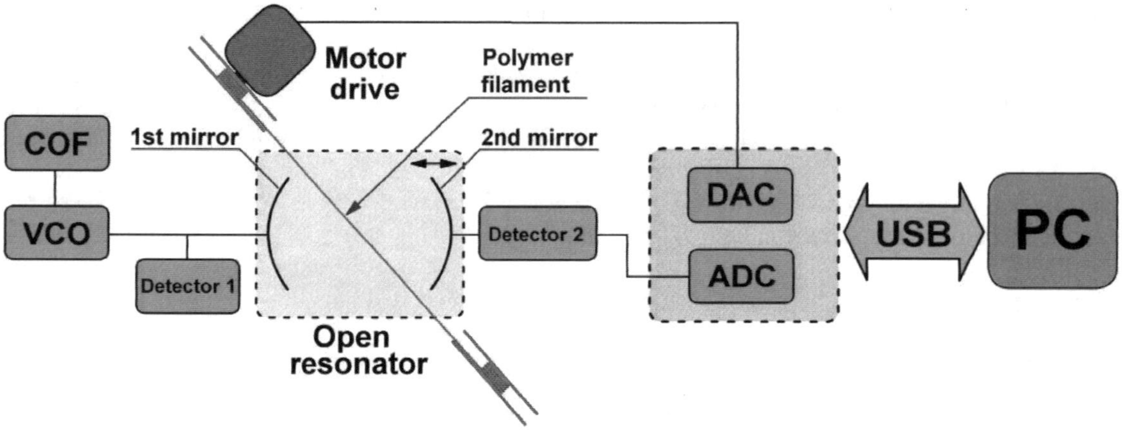

Figure 2. The experimental setup of quasi-optical resonant single-frequency flaw detector for heterogeneity of polymer filament for 3D printing.

The device has two detectors - for the reference signal excited directly by the generator, as well as the signal that passed through the resonator. Since the device is designed for a dynamic mode, in order to reduce the influence of displacements and vibrations of the filament when it is pulled, the conductor is inclined in the plane of the electric field in an open resonator. The signals from the detectors are recorded with a 12-bit analog-to-digital converter (ADC). The filament pulling mechanism is controlled by a 8-bit digital-to-analog converter (DAC) through a PC via the USB bus.

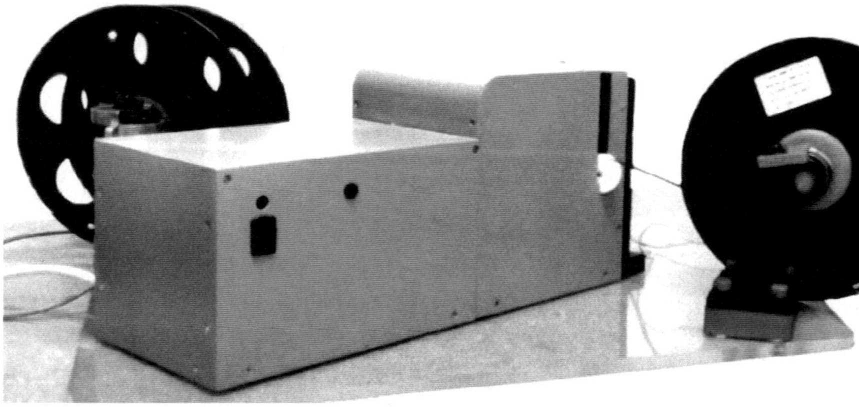

Figure 3. Appearance of the quasi-optical single-frequency flaw detector of heterogeneities of polymer filaments and a filament broaching system.

The open measuring resonator is placed in a protective shield (Figure 3), while the filament can be freely introduced and stretched through a sufficiently wide (about 1 cm) gap. The diameter of the mirrors of the open resonator was 160 mm; the radius of curvature was 200 mm.

4. Approbation

For the approbation of quasi-optical resonant single-frequency flaw detector 5 samples (Table 1) of industrial polymer filaments, polyethylene terephthalate glycol (PET-G), polystyrene (HIPS), acrylonitrile styrene acrylate (ASA), conductive acrylonitrile butadiene styrene, conductive polylactide were selected. By the coaxial method at frequency of 10 GHz real (ε') and imaginary (ε'') parts of the permittivity of printed test samples by additive technology were measured.

Table 1. Parameters of the filaments and signal level on detector 2 of quasi-optical resonant single-frequency flaw detector at maximum (V_{det}(Pmax)) of resonance curve.

Sample	$D_{filament}$ (mm)	V_{det}(Pmax) (mV)	ε' (f=10 GHz) (rel. units)	ε'' (f=10 GHz) (rel. units)
Empty resonator		1309.4		
Polyethylene terephthalate glycol	1.770	1276.3	2.569	0.01866
Polystyrene	1.765	1267.2	2.253	0.00363
Acrylonitrile styrene acrylate	1.713	1256.9	2.429	0.02795

As follows from the table, the value of the transmission coefficient of the resonator at the resonant frequency is ambiguously associated with losses in the filament. This is possible in a certain region where, due to a decrease in losses in the volume of an object (ohmic or thermal losses) with a decrease in the tangential component of the electric field, provided that there are small diffraction losses of the resonant beam on the filament [19]. With the exception of this region, in other cases, the Q factor of the cavity decreases with increasing conductivity of the sample. Thus, as for an ultrathin conductor, for a polymeric fiber, it is also necessary to take into account the ambiguity, which consists in the fact that the same losses recorded by the resonator sensor can correspond to different fiber characteristics. This is especially true for a single-frequency sensor.

Table 2. Parameters of the conductive filaments and signal level on detector 2 of quasi-optical resonant single-frequency flaw detector at maximum (V_{det}(Pmax)) of resonance curve.

Sample	$D_{filament}$ (mm)	V_{det}(Pmax) (mV)
Acrylonitrile butadiene styrene (conductive)	1.740	282.8
Polylactide (conductive)	1.730	5.8

As follows from table 2, an increase in the conductivity of the filament leads to a significant decrease in the quality factor of the resonator, which reduces the resolution of the sensor. Thus, the proposed method is most optimal for filaments with close values of the parameters (complex permittivity) and allows you to record small deviations in the characteristics of the filament. The deviation of the resonance curve from the initial value should not exceed half its width.

Conclusion

Thus, the possibility of using a quasi-optical resonator for flaw detection of heterogeneities of polymer composite for a 3D printer is shown. This system can be used in the production of filament for 3D-printers working on the fused deposition method, for the purpose of non-contact rapid analysis of the homogeneity of the manufactured product.

References

[1] Jaksic N I and Desai P D 2018 Characterization of resistors created by fused filament fabrication using electrically-conductive filament *Procedia Manuf.* **17** 37–44

[2] Kim K, Park J, Suh J hoon, Kim M, Jeong Y and Park I 2017 3D printing of multiaxial force sensors using carbon nanotube (CNT)/thermoplastic polyurethane (TPU) filaments *Sensors Actuators, A Phys.* **263** 493–500

[3] Flowers P F, Reyes C, Ye S, Kim M J and Wiley B J 2017 3D printing electronic components and circuits with conductive thermoplastic filament *Addit. Manuf.* **18** 156–63

[4] Chen Q, Chen X and Xu K 2017 3-D printed Fabry–Perot resonator antenna with paraboloid-shape superstrate for wide gain bandwidth *Appl. Sci.* **7** 1134

[5] Kuleshov G E, Zhuravlyova Y V. and Dotsenko O A 2015 Electromagnetic response from composite radiomaterials based on multiwall carbon nanotubes at microwave frequencies *2015 Int. Sib. Conf. Control Commun. SIBCON 2015 - Proc.* 0–4

[6] Kuleshov G E, Zhuravleva Y V. and Chernobrova D A 2014 Microwave electromagnetic parameters of composite radiomaterials based on carbon nanostructures and ferrimagnetics *Int. Conf. Young Spec. Micro/Nanotechnologies Electron Devices, EDM* 158–60

[7] Kuleshov G E and Badin A V 2019 Study of electromagnetic characteristics of polymer materials based on single-walled and multi-walled carbon nanotubes *IOP Conf. Ser. Mater. Sci. Eng.* **525** 012030

[8] Bastola A K, Paudel M and Li L 2018 Development of hybrid magnetorheological elastomers by 3D printing *Polymer (Guildf)* **149** 213–28

[9] Yang F, Zhang X, Guo Z, Ye S, Sui Y and Volinsky A A 2019 3D printing of NdFeB bonded magnets with $SrFe_{12}O_{19}$ addition *J. Alloys Compd.* **779** 900–7

[10] Wang X, Jiang M, Zhou Z, Gou J and Hui D 2017 3D printing of polymer matrix composites: A review and prospective *Compos. Part B Eng.* **110** 442–58

[11] Lei Z, Chen Z, Zhou Y, Liu Y, Xu J, Wang D, Shen Y, Feng W, Zhang Z and Chen H 2019 Novel electrically conductive composite filaments based on Ag/saturated polyester/polyvinyl butyral for 3D-printing circuits *Compos. Sci. Technol.* **180** 44–50

[12] Fantino E, Chiappone A, Roppolo I, Manfredi D, Bongiovanni R, Pirri C F and Calignano F 2016 3D printing of conductive complex structures with in situ generation of silver nanoparticles *Adv. Mater.* **28** 3712–7

[13] Sezer H K and Eren O 2019 FDM 3D printing of MWCNT re-inforced ABS nano-composite parts with enhanced mechanical and electrical properties *J. Manuf. Process.* **37** 339–47

[14] Leigh S J, Bradley R J, Purssell C P, Billson D R and Hutchins D a. 2012 A simple, low-cost conductive composite material for 3D printing of electronic sensors *PLoS One* **7** 1–6

[15] Dudorov S N and Lioubtchenko D V Open resonator technique for measuring dielectric properties of 3–5

[16] Badin A V, Bessonov V V, Dorozhkin K V, Dorofeev I O, Hiu L B and Dunaevskii G E 2018 Terahertz resonator diagnostics of filamentary dielectric objects *proc. in 43th International Conference on Infrared, Millimeter, and Terahertz Waves (IRMMW-THz 2018) 1-2*

[17] Badin A V. and Dunaevskii G E 2014 Optimization of resonator for local quasi-optical SHF and EHF diagnostics of inhomogenious materials anisotropy *2014 24th International Crimean Conference Microwave & Telecommunication Technology* (Sevastopol: IEEE) pp 910–1

[18] Komiyama B, Kiyokawa M and Matsui T 1991 Open resonator for precision dielectric measurements in the 100 GHz band *IEEE Trans. Microw. Theory Tech.* **39** 1792–6

[19] Dorofeev I O and Dunaevskii G E 2012 Losses in an open microwave resonator with an ultrathin cylinder **54** 53–9

Increasing Fatigue LIfe of 09Mn2Si Steel by means of High-Temperature Multistep Helical Rolling

I V Vlasov[1], N S Surikova[1], S V Panin[1,2], A V Yakovlev[2] and J F Gomorova[1]

[1]Institute of Strength Physics and Materials Science SB RAS, 2/4, pr. Akademicheskii, Tomsk, 634055, Russia
[2]National Research Tomsk Polytechnic University, 30, Lenin Avenue, Tomsk, 634050, Russia

E-mail: viv@ispms.tsc.ru

Abstract. The effect of high temperature helical rolling (HR) on structure and fatigue life of 09Mn2Si pipe steel has been studied. With the use of transmission electron microscopy there was revealed that rolling gives rise to refinement of ferrite grains and cracking (fracturing) of cementite plates within the pearlite phase. The effect manifests itself to the greatest extent in the surface layer where due to the rolling the level of plastic deformation was the highest. Data of microhardness measurements confirms the gradient pattern of strain hardening over the cross section during the HR occurs while the most intensive microhardness increasing take place at the depth of up to 3 mm. According to the mechanical testing results the helical rolling of 09Mn2Si steel gives rise to increasing the level of deforming stress at the yield plateau as well as the proportionality limit with a general decrease in the relative elongation. At the same time, despite the strain hardening resulting from the helical rolling the mechanisms of plastic deformation which manifest themselves in the form of parabolic hardening with a smooth decrease in the flow stress level after neck formation are preserved in the steel. During the cyclic tension the number of cycles prior to failure increases from 2.5 to 3.8 times that depends on the location of specimens' cutting from the rolled rod. The highest improvement in fatigue fracture resistance is registered for specimens cut out from the core of the rolled rods.

1. Introduction

Development of new technologies sets new requirements to the properties of structural materials. One of the effective methods for increasing the strength properties of already developed steels is the formation of certain structural-phase states in the bulk material [1, 2]. Their variation at the processing stage allows one to obtain the required values of hardness, strength, ductility, and crack resistance. Most often such treatment includes thermal component for the formation of desired phases and subsequent plastic deformation for the purpose of additional strain hardening. Low-alloyed low-carbon ferrite-pearlite steels are most treatable with these types of processing. Violation of processing conditions or improperly selected temperature parameters can give rise to a highly stressed state usually accompanied by embrittlement of material. In this regard, it is relevant to find out complex thermal-mechanical treatments in order to attain high strength while maintaining sufficient ductility. For this purpose, high-temperature helical rolling can be utilized, which makes it possible to exert a controlled multistage thermal-mechanical impact on the material processed [3].

The main point of high-temperature multistep helical rolling is a gradient-like structure modification in the bulk material [4]. In doing so, the macroscopic structure is formed ensuring increased strength in the surface layer and maintains high ductility in the core. Preliminary established temperature and strain of helical rolling regimes [5] (including the number of passes) allow one to form a multi-level structure with high dissipative properties in the bulk material. This makes it possible to effectively suppress plastic flow localization and significantly hinder nucleation and propagation of cracks.

Low-alloyed structural 09Mn2Si steel which is widely used both at constructing pipelines as well as for manufacturing structural components was taken as an object of study. The paper is aimed at research of the impact of the helical rolling (HR) on the structure and fatigue properties of this steel.

2. Experimental

High-temperature multistep helical rolling of steel with gradual decreasing rolling temperature at each subsequent pass was performed with the help of a three-roll laboratory scale rolling machine RSP "14-40" in the temperature range 1123-773 K (850-500 °C). After each rolling pass the rod was quenched in cold water. The total degree of true logarithmic strain of the steel during the rolling was 0.8 (relative strain 73.93 %).

Dog-bone shape specimens with a gauge length of $16 \times 3 \times 1$ mm were cut out from the rods along the longitudinal direction by the electro-spark machine. Static tension tests were carried out with the use of an Instron 5582 electromechanical testing machine; the tensile loading rate was 0.6 mm/min.

The fatigue tests were conducted with the use of Biss UTM 150 servo-hydraulic frame. In the paper, for the crack growth diagrams as well as photographs depicting their development the results are given being most close to the average ones for each group. The fatigue ratio R was set at 0.1 with the maximum load per cycle of 290 MPa and the loading frequency of 20 Hz. To localize the processes of fatigue crack nucleation, the I-shaped notch with the length of 400 μm and a tip radius of 125 μm was introduced in the middle part of each specimen.

The size of ferrite grains was measured according to ASTM E 112-96. Microhardness measurements were conducted with the help of a PMT-3 instrument with the load on the Vickers pyramid of 0.98 N (100 g). Fine crystalline structure was examined over thin foils with the use of a transmission electron microscope CM-12 (Philips) at the accelerating voltage of 200 kV.

3. Results

3.1. *Metallography.* As-received 09Mn2Si steel possesses a ferrite-pearlite structure (Figure 1, a). After the rolling, a finely dispersed structure consisting of ferrite as well as fractured plates of the pearlite phase is seen in the surface layer both along the longitudinal and transverse directions (Figure 1, b). At the depth above 1 mm in a longitudinal direction, some individual ferrite grains become visible (Figure 1, c). They have been kept undamaged by the rolling induced deformation and are elongated towards the longitudinal direction with a slight deviation to the center of the rod.

At observation in the transverse direction such grains appear as equiaxed light phases, sometimes slightly elongated (due to lateral deviations). With approaching to the rod center an increase of the size and the number of distinguishable elongated ferrite grains becomes evident, while at the distance of 4 mm their texture becomes clearly visible (Figure 1, d).

At this depth (at observation in the transverse direction) both the ferrite and pearlite phases become distinguishable. However, the helical rolling induced deformation governs poor appearance of pearlite interlayers located between ferrite grains. However, in the as-received state steel these phase boundaries look more “ragged” (caused by mixing of phases during the helical rolling).

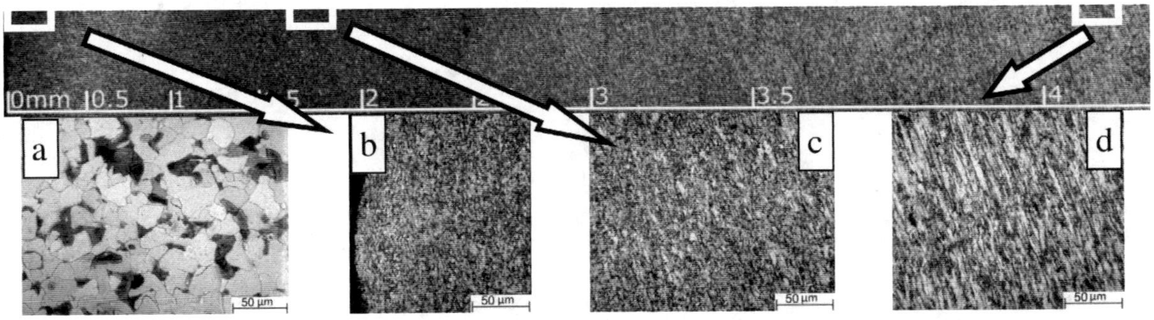

Figure 1. Optical images of the microstructure of (a) as-received steel specimen; and (b-d) after the helical rolling: longitudinal section.

3.2. *Transmission electron microscopy.* With the help of the transmission electron microscope the fine structure of the as-received steel as well as one after the HR have been studied (Figure 1, a, b). Characteristic cementite plates with a thickness of 100-150 nm are seen within the pearlite grain of the steel not subjected to the rolling. There are carbide inclusions with the size of 250-400 nm within the ferrite phase (Figure 1, a). During the rolling, refinement of ferrite grains and fracture of cementite plates has occurred (Figure 1, b). The average subgrain size of the rolled steel is about 300 nm, while in the surface layer it can reach 100 nm.

3.3. *Mechanical properties.* The microhardness over the cross section of the rod after the HR was measured (Figure 1, c). The highest degree of the microhardness increase is registered in the surface layer up to 3 mm depth, while the hardening occurs throughout the entire bulk of the rod.
Specimens for static stretching were cut out from the HR-processed rod at the depth of 3 mm where the the microhardness was lower than in the surface layer.

At the loading diagram of the as-received material one can easily distinguish a yield tooth and a yield plateau being characteristic for low carbon steels (Figure 1, d, table 1). The diagram exhibits a long stage of parabolic hardening and high value of relative elongation at break. Despite the increase in the microhardness, at the loading diagram for the HR-treated steel both the yield tooth and the yield plateau present indicating that the strain hardening mechanisms (which is of particular importance under impact loading) are kept. At the same time, the stresses at the yield point as well as the proportionality limit have increased by 26 %. Also, the stress at the yield tooth is comparable with the value of tensile strength. The duration of the yield plateau has increased by 3 times.

Table 1. Results of static tension tests.

State	Proportionality limit (MPa)	Flow stress at yield plateau (MPa)	Upper yield point (MPa)	Ultimate strength (MPa)	ε (%)
Initial state	163±10	297±10	307±20	443±20	32±3
After HR	219±10	403±20	431±20	438±20	21±3

The fracture of the HR-treated steel specimens after the neck formation occurs smoothly, much like to as-receive steel. The fact of keeping plasticity mechanisms at the yield plateau stage and the smooth failure pattern after the neck formation (with quite manifested strain hardening) can be explained by the decrease of the grain size and texture formation during the rolling. The newly formed extended grain boundaries prevent the plastic deformation development, while the ferrite grains that avoid significant hardening favouring accumulation of defects. That is why they are responsible for presence of the yield tooth and plateau at the loading diagram. It should also be noted that the hardening of the material after the helical rolling still gives rise to the decrease of the relative elongation value (by 34 %).

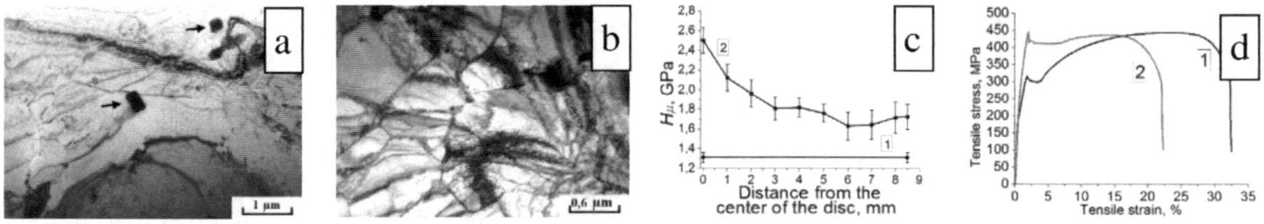

Figure 2. TEM-microimages of fine structure a) as-received (1); b) after the HR (2); c) cross-section microhardness graph in the rod as a function of distance from the surface; d) loading diagram.

Fatigue tests were carried out for the as-received 09Mn2Si steel specimens and ones after the HR. Since the rolled rod had a gradient structure over the cross section the specimens were cut out from three different locations of the rod's cross section (Figure 3, a).

It was revealed that the high-temperature helical rolling gives rise to increasing fatigue life at least by 2.5 times (table 2). Depending on the location of the specimens cutting the number of cycles prior to failure changes. The lowest increase in fatigue life after the HR is observed for specimens cut out from the area 1 where microhardness has increased to the maximum extent. Specimens from the core of the rod being characterized by the lower microhardness (as compared with specimen from the area 1) showed the largest number of cycles prior to failure. As compared to the as-received steel the number of cycles prior to fracture for the area 3 specimens have increased by 3.8 times.

Based on the results of a quantitative image analysis of the specimen surface taken during cyclic testing some graphs of the main crack opening were constructed (Figure 3, b). The calculation of the crack opening value was initiated from the moment when its length reached 80 µm (to provide the reliability of the measurement results at the used magnification of the optical surface photo observation). One can see intensive crack growth over the entire measurement range in the as-received steel that is associated with a higher ductility and lower strength of the material. For the HR case the intensive crack growth is observed mainly at the final stage of cyclic loading ($N/N_{failure}$> 90 %).

Also, for the aim of quantitative comparison of fatigue fracture regularities for various specimens the graphs of their elongation during cyclic loading were constructed (Figure 3, c). It is seen that the lowest elongation is typical for a specimen cut out from area 1. The elongation is slightly greater for a specimen taken from area 2, while the largest one is typical for the specimens cut out from area 3 which is comparable to the elongation for the as-received steel (curve 4).

Thus, the key effects identified here are: 1) in terms of the current number of cycles normalized over the total one prior failure, a macrocrack nucleates at approximately the same time; although in absolute terms the HR significantly delays the beginning of the main crack propagation stage. 2) crack opening in the as-received steel occurs at significantly higher velocity as compared to the HR specimens cut from locations 1–3; 3) the value of elongation during cyclic loading as well as the velocity of this process in the specimen cut out from zone 3 are almost identical. This indicates a high ductility of the formed structural-phase state combined with high strength properties.

The image analysis of the specimen's surface during cyclic loading (Figure 4) was carried out. The number of cycles was determined for the instance when a main crack nucleates. It can be seen that for the as-received steel (Figure 4, a-d), the value of $\boldsymbol{N_{cr.init.}/N_{fail}}$ is equal to 30 % (table 2). This is characteristic of ductile steels when the duration of the main crack propagation stage significantly exceeds that of its nucleation. After the main crack has formed a manifested contraction in the notch region is seen (Figure 4, b). Subsequently, in the front of the crack tip a two-pestle like plastic deformation zone being traditional for ductile materials is formed (Figure 4, c). Its size increases with the crack growth, and is decorated with a pronounced initial grain structure of the material. The crack branches up during growth and plastic deformation, especially in the last stages, manifests itself to the greatest extent (Figure 4, d).

The specimens after the HR taken from location 1 possess the lowest value (20%) of the $N_{cr.init.}/N_{fail}$ parameter (table 2). By the authors' opinion the pointed result is associated with a high degree of strain hardening of the material in this region since due to significant defect density the crack nucleates quite early (but only with respect to the normalized number of cycles prior failure).

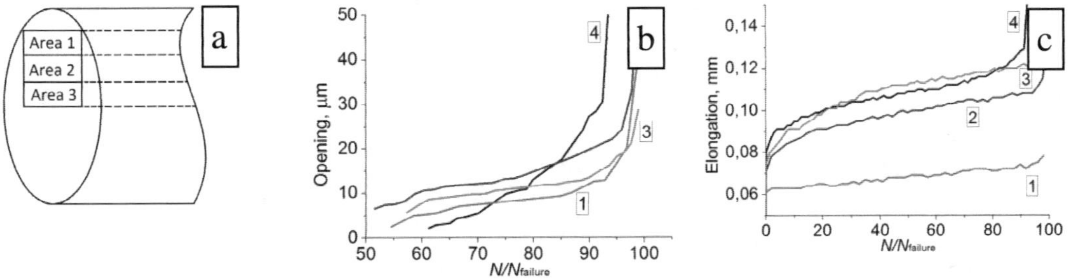

Figure 3. a) scheme of the rod and locations for cutting out specimens; b) crack opening graph; c) elongation during cyclic loading; 1) after the HR, area 1; 2) area 2; 3) area 3; 4) as-received state.

Table 2. Results of fatigue tests.

State	Area	The number of cycles to failure	$N_{cr.init.}/N_{fail}$ (%)
Initial state	-	80 000±5 000	30±4
	Area 1	200 000±10 000	20±2
After HR	Area 2	225 000±12 000	30±3
	Area 3	305 000±15 000	45±5

After the nucleation of the main fatigue crack (Figure 4, f) there are no pronounced traces of contraction, while the crack grows without branching and with minimal evidences of plastic deformation development (Figure 4, g). At the final stage, some step-like lines are formed near the main crack, being oriented towards the direction of its growth (Figure 4, h) thereby forming slight contraction. However, in comparison with as-received steel much more brittle fracture pattern is observed, while plastic deformation at the final stage is slightly pronounced. According to the authors opinion, this effect is associated with the formation of finely dispersed grain structure during the HR, when numerous boundaries of the structural elements occur impeding the main crack growth.

The highest value of the parameter $N_{cr.init.}/N_{fail}$ is observed in the specimens cut from the location 3 which is equal to 45 %. This ratio is more typical for less ductile materials, and more than 2 times differs from that for higher strength specimens cut out from the zone 1. There are no evidences of plastic contraction at crack initiation. The main crack branches during its propagation and weak traces of plastic deformation development are visible near its tip. At the final stage of crack growth slight contraction is also registered. According to the authors' opinion, the maximum increase in the fatigue life of specimens cut out from zone 3, is caused by the following reason. During the HR, quasi-laminate (textured) structure is formed in zone 3. In doing so, a fatigue crack growths towards the direction of the preferred orientation of its structural elements. In this case, the density of defects formed during the HR as compared to zone 1 is lower (as evidenced from the microhardness measurements).

IOP Conf. Series: Materials Science and Engineering **731** (2020) 012019 doi:10.1088/1757-899X/731/1/012019

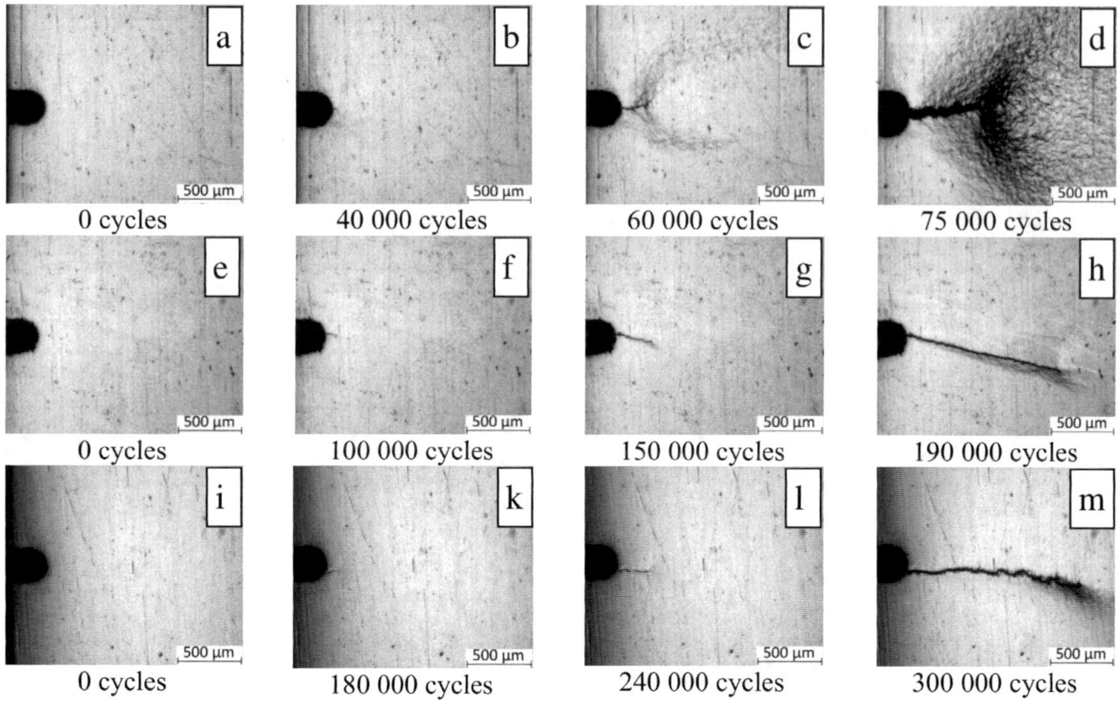

Figure 4. Images of specimens surface under cyclic tension; a) - d) as-received state; e) - h) after the helical rolling (area 1); i) - m) after the helical rolling (area 3).

As a result of the strength increase for specimens taken from zone 3, the parameter $N_{cr.init.}/N_{fail}$ has a maximum value (by the way, it gradually increases when moving from zone 1 (20 %) to the zone 2 (30 % which corresponds to the as-received steel) and then to the zone 3 (45%)). Thus, from the optimality point of view (of the gradient structure formation over the cross section), it is the material in zone 3 that should be considered as one having the best structural–phase state.

Conclusion

The structure and mechanical properties of 09Mn2Si steel were studied after the high-temperature five-step HR. It was revealed that the rod treatment results in finely dispersed structure formation in the surface layer up to 3 mm deep being caused by intensive plastic deformation. This is accompanied with fibrous-like textured structure formation consisting of thin ferrite grains elongated towards the rolling direction in the underlying layers. The greatest increase in microhardness is observed in the surface layer at the depth of up to 3 mm.

Under static tension in the HR-specimens the stress at the yield plateau as well as the proportionality limit increase by 26 %, while the tooth yield stress is comparable to the tensile strength. The duration of the yield plateau increases by 3 times. The i) preservation of the “parabolic hardening” stage, yield tooth and plateau at the tensile diagram, ii) the high level of deforming stresses, iii) the maintenance of plastic deformation mechanisms and iv) the smooth flow stress decreasing after the “neck” formation with simultaneous strain hardening are logically explained by the decrease in grain size and texture formation during the rolling including formation of numerous newly formed extended grain boundaries.

The structure of specimens cut out from various location of the rod after the helical rolling has been analyzed. It was found that the specimens taken from the core of the rod (area 3) are characterized by a more equilibrium and less deformed state that is manifested in lower microhardness, higher elongation under cyclic loading being comparable with as-received steel and more ductile pattern of the main fatigue crack growth (in comparison with area 1). Apparently, the preservation of high ductility and the presence of the fibrous-like texture being oriented normally the direction of the crack propagation make it possible to increase the duration of main crack propagation stage while due to the increase in strength the number of cycles required for the critical accumulation of defects (the stage of the fatigue crack initiation) increases. This results in a maximum increase in fatigue life up to 3.8 times.

On the other hand, in specimens cut out from the area 1 due to the most intensive straining during the HR the highly defective structure having a lower ductility is formed. This is manifested in minimal elongation of specimens under cyclic loading and a more brittle pattern of the main crack growth. This gives rise to earlier crack nucleation ($\boldsymbol{N}_{\mathbf{cr.init.}}/\boldsymbol{N}_{\mathbf{fail}}$ =20%) and a lower number of cycles prior to failure. However, due to the strain hardening, the number of cycles prior to failure for specimens of this type remains 2.5 times higher as compared to as-received steel.

Acknowledgments
The study was performed in the framework of fundamental research Program of the Russian State Academies of Sciences for 2013–2020, line of research III.23 as well as a partial support of RFBR Grants № 18-38-00679 and 18-08-00516. Financial support of RF President grants i) Scholarship SP-2456.2019.1 and ii) RF President Council Grant for the support of leading research schools NSh-5875.2018.8 is also acknowledged.

References
[1] S S Samant, I V Singh and R N Singh 2018 *Materials Science and Engineering: A* **738** 135-152
[2] Nagaraj M and Ravisankar B 2018 *Materials Science and Engineering: A* **738** 420-429
[3] Naizabekov A, Lezhnev S, Arbuz A and Panin E 2013 *Metallurgical Research & Technology* **115**(2) 213
[4] Derevyagina L S, Gordienko A I and Kashiro P O 2019 *Russian Physics Journal* **61**(11) 1971–1977
[5] Derevyagina L S, Gordienko A I, Pochivalov Yu I and Smirnova A S 2018 *Physics of metal and mettallography* **119**(1) 83-91

Cytotoxic Effects of Granulated Hydroxyapatite with Various Particle Size Distribution

O Zakharova[1,2], A Gusev[1,2], K Chuprunov[1], A Yudin[1] and D Kuznetsov[1]

[1]Department of Functional Nanosystems and Hugh Temperature Materials, National University of Science and Technology «MISIS», Moscow 119991, Russia
[2]Research Institute for Environmental Science and Biotechnology, Derzhavin Tambov State University, Tambov 392000, Russia

E-mail: olgazakharova1@mail.ru

Abstract. Hydroxyapatite-based materials show promise in such applications as reconstructive surgery, stomatology and cosmetology, they can also be used as components of toothpaste, disinfectant products and dietary supplements (DS). However, cytotoxicity of hydroxyapatite powders synthesized via various methods towards a variety of biological cells requires a thorough study.
Cytotoxicity of hydroxyapatite powder samples with spherical particles in three size ranges: 5-25 μm, 25-45 μm and 40-125 μm, synthesized by hydrothermal synthesis with further ultrasonic and spray-drying treatment, was studied on human blood cells and on gram-negative bacteria *E.coli*.
The examined hydroxyapatite samples have no toxic effect on human blood leukocytes (in concentration 40 mg per 1 ml H_2O or normal saline) and on gram-negative bacteria *E.coli.* (in concentration 40 mg per 1 ml H_2O or normal saline) regardless of the contact time and particle size. The obtained data demonstrate safety of the examined materials and absence of toxic effects towards the test objects. This allows us to regard the studied hydroxyapatite samples as biocompatible.
The present study results could be used to develop products and medicinal drugs based on hydroxyapatite granules synthesized by means of hydrothermal synthesis technique with further sonication and spray drying.

1. Introduction

Nanostructured apatite-based coatings are widely used on metal implants to improve their biocompatibility, advance osteointegration, promote bone fixation and accelerate bone growth. Development of new biocompatible and nontoxic materials with possibility of achieving direct osteointegration is an important task. Hydroxyapatite ($Ca_{10}(PO_4)_6(OH)_2$) is a promising material for performing the presented task. The hydroxyapatite-based could be successfully employed in reconstructive surgery (diseases and injuries of the human musculoskeletal system, maxillofacial surgery, filling damaged and diseased bones), stomatology, as components of toothpaste, cosmetological and disinfectant products and as dietary supplements [1-3] due to their high biocompatibility, wear resistance and non-toxicity. Hydroxyapatite-based materials are readily available, inexpensive and easy-to-use. Hydroxyapatite, as one of the major inorganic components of human bones and teeth, is widely studied as a filler and implant coating material [2].

In addition, hydroxyapatites have also garnered attention in soft tissue regeneration due to their excellent biocompatibility with soft tissues such as skin, muscle, and other. Numerous studies have shown that hydroxyapatite can activate fibroblasts and accumulate vessel endothelial cells and thereby support the healing of skin wounds [4].

Hydroxyapatite bioceramics have sealing contact and strong adhesion with skin tissue to prevent exit-site and tunnel bacterial infection, which suggests that hydroxyapatite-based materials might be utilized as percutaneous devices [5]. Composite products with a hydroxyapatite component have been successfully developed for soft-tissue augmentation [6], and a study has demonstrated that hydroxyapatite nanoparticles could stimulate the axonal outgrowth, suggesting that hydroxyapatite might provide a new approach for therapy or prevention of nerve injury [7].

Biocompatibility examination is the most important step in developing new biomaterials. Two major aspects have to be taken into consideration while performing biocompatibility tests on novel medicinal materials, i.e. the material can produce toxic effects on the cells in immediate contact and on the body as a whole; and the material itself can undergo degradation or lose its functionality under the influence of the internal environment.

The cytotoxicity test allows evaluation of the toxic potential of the material at the cellular level, i.e. *in vitro*. The method provides quantitative and comparable results, therefore, it allows one to quickly evaluate and select toxic materials before conducting animal studies. Either cell cultures or unicellular microorganisms (bacteria, microalgae) are usually used as test objects in assessing cytotoxicity. Initially the test objects' survival following treatment with the material is assessed, and in the event of any negative effects the possible toxicity mechanisms are established.

This paper shows the results of cytotoxicity evaluation performed for spherical hydroxyapatite samples obtained via hydrothermal synthesis method with ultrasonic treatment and spray drying, human blood cells and gram-negative bacteria *E.coli* are used as test objects.

2. Materials and methods

The hydroxyapatite samples were synthesized by the hydrothermal synthesis method. Calcium nitrate tetrahydrate ($Ca(NO_3)_2$, AR Grade CAS 13477-34-4), diammonium hydrogen phosphate $(NH_4)_2HPO_4$, AR Grade) ammonium hydroxide (NH_4OH, AR Grade, CAS 7664-41-7) produced by Reachem Co (Russia) were used as precursors in reaction (1).

$$10Ca(NO_3)_2 + 6(NH_4)_2HPO_4 + 8NH_4OH \rightarrow Ca_{10}(PO_4)_6(OH)_2 + 20NH_4NO_3 \quad (1)$$

The resulting suspension was mixed by a mechanical mixer Heidolph RZR 2051. The Ca/P ratio was calculated as 1.67. The suspensions were sprayed with the spray dryer equipment Buchi Mini Spray Dryer B-290 with the ultrasonic atomizer Buchi B-290 Ultrasonic Package. As a result we obtained granulated hydroxyapatite samples with different particle size distributions (5 – 25 μm, 25 – 45 μm and 40 – 125 μm), depending on the spray drying process parameters.

This approach provides one with a possibility to control solubility of materials used for implant manufacturing from rapidly soluble to insoluble (calcium tetraphosphate > α-tricalcium phosphate dehydrate > anhydrous dicalcium phosphate > octocalcium phosphate tricalcium phosphate > hydroxyapatite) via controlling the ratio between hydroxyapatite and calcium phosphate substances.

The cytotoxicity analysis was performed on human blood cells (mononuclear leukocytes) and gram-negative bacteria *E.coli*.

The intact donor blood (n = 6) stabilized with 25 unit/ml of heparin was suspended with sterile normal saline (isotonic sodium chloride solution, 0.9 % solution of NaCl, PanEco, Russian Federation), in a 1:2 ratio. The mononuclear leukocytes (ML) were separated with the one-step ficoll gradient (PanEco, Russian Federation, density 1.077g/cm^3) with centrifugation at 400 g for 30 minutes. The diluted blood was carefully layered on a tube wall with a pipette above 5 cm^3 of ficoll – verographin poured into centrifuge tubes (V = 50 ml).

The mononuclear leukocytes forming the interphase ring were collected with a pipette and triply washed with Roswell Park Memorial Institute medium (RPMI - 1640). The nutrient medium was prepared on the RPMI-1640 basis according to the following procedure: 10 % of fetal calf serum, 2 mM of glutamine and 5000 ME/ml of streptomycin and penicillin at 5000 ME/ml were added to

RPMI-1640 under sterile conditions. After the final rinsing from ficoll the samples were resuspended in the nutrient-rich medium.

The cell viability in the suspensions was analyzed by the trypan blue dye exclusion method using optical microscopy. The cell concentration was calculated in the Goryaev chamber. The live mononuclear leukocytes concentration in the biological medium corresponded to 3.2 – 3.4 • 10^6 cells per 1 ml of the medium.

The resulting suspension was added, to the wells of a standard plate, 1 ml per well, containing 40 mg of hydroxyapatite, gently stirred and incubated for 30 min at 37 °C in a reciprocating shaker (ELMI ST–3L, RF) at 350 rpm. The control leukocytes suspension was incubated under the similar conditions, with no hydroxyapatite added.

The cytotoxicity analysis was carried out using the colorimetric assay for assessing cell metabolic activity (MTT colorimetric assay). After incubation 100 μl of 5 % MTT solution in normal saline was added to each well with further incubation for 4 h in a CO_2 incubator at 37 °C and CO 5 %.

The volume of 800 μl of the suspension was moved from each well to 2 ml centrifugal microtest-tubes (Eppendorf) and centrifuged for 10 h at 1800 rpm/min in an Eppendorf S 415 R centrifuge. The supernatant fluid was carefully removed with a pipet leaving the cell precipitate in the tubes.

Dimethyl sulfoxide (DMSO) was added to every test tube (500 μl/tube). The samples were incubated in a rotary shaker at room temperature for 30 minutes at 350 rpm/min. When the formazan was diluted, the dyed DMSO solution was pipetted into the wells of a 96-well plate (100μl/well). The effect of hydroxyapatite on the cells from each donor was determined via characterization of the cell lysate in parallel wells. The intensity of purple staining was determined by characterization of the optical density of formazan solution in DMSO using a Multiskan MS spectrophotometer (Labsystem, Finland) at 540 nm wavelength.

The index cytotoxicity (IC) of the hydroxyapatite samples was calculated using the equation (2)

$$IC, \% = 100 - (OAs/OAc \cdot 100) \qquad (2)$$

where OAs – optical absorption of the obtained hydroxyapatite sample, OAc – optical absorption of the intact control.

The cytotoxicity analysis on bacterial cells was performed according to the bioluminescence method [8-10] on *Esherichia coli* with the cloned *luxCDAEB* genes of *Photobacterium leiognathi*. The toxic effect produced by the studied nanomaterial on the bacteria was determined by inhibition of their bioluminescence. The measurements were performed on the specialized luminometer Biotox-10 (Russian Federation). The first stage of the test was performed immediately after the sample preparation and the second stage was performed after a 24-hour ageing period.

The lyophilized bacterial culture was revived in distilled water and in normal saline (pH = 7.0 – 7.4).

The hydroxyapatite suspensions were also prepared by dispersion in distilled water and in normal saline (pH = 7.1 ± 0.2). The quantity of 400 mg from each hydroxyapatite sample was accurately weighed on an analytical balance ViBRA HT (Shinko Denshi, Japan) with the accuracy ± 0.0001 g, poured into a 10 ml glass tube with the liquid medium and mixed by intensive shaking with further ultrasonic treatment.

The study was performed at different exposition times: 5 min, 30 min, 60 min and 120 min.

The descriptive primary analysis data are presented in the format sample mean (χ) ± standard deviation value (SD). Statistical analysis was carried out using Statistica 6.0, StatSoft software with the "Basic statistic/Tables" and "Non-parametric data" modules.

3. Results and discussion

The cytotoxicity index of the hydroxyapatite samples was calculated for each donor, based on the optical absorption values obtained after measuring formazan in DMSO solution extinction both for the test samples and control samples. The description analysis of the obtained primary data generalising the cytotoxicity index values of the tested and control samples was performed employing the statistical analysis software 6.0 with module "Basic statistic/Tables" (figure 1).

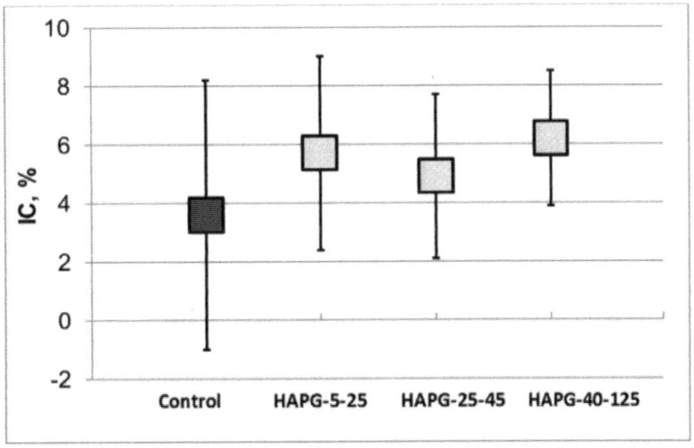

Figure 1. IC value distribution in the control before and after co-incubation with the hydroxyapatite samples in mononuclear leukocytes suspensions.

The presented data show that the leukocytes death intensity induced by the hydroxyapatite samples never exceeded 9 % for all the tested donors. Moreover, the statistical comparison analysis of the IC indicators in the control group, where the leucocytes were subjected only to constant agitation in a shaker, and in the experimental group, where the leukocytes were in contact with the hydroxyapatite particles in addition to physical agitation revealed no significant differences.

The hydroxyapatite cytotoxicity analysis in an aqueous solution in relation to *E.coli* bacteria was carried out after 5, 30, 60 and 120 min exposition. The observed results are presented in figure 2.

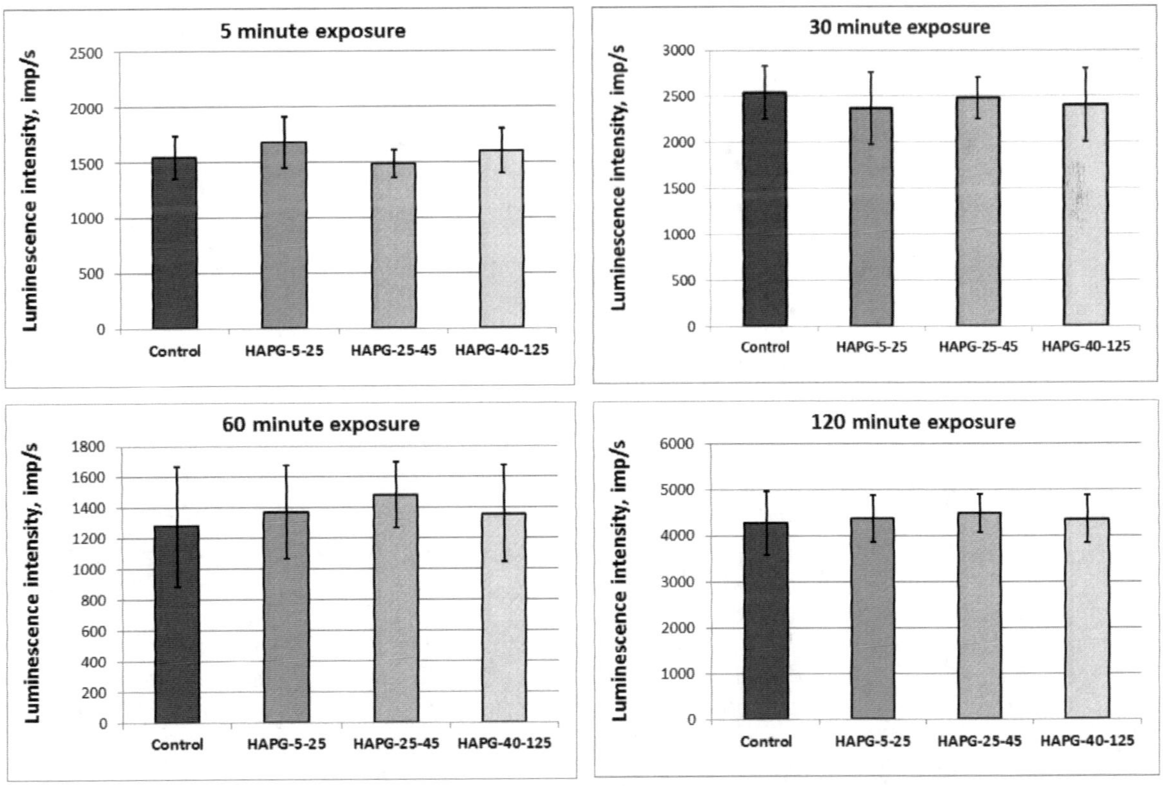

Figure 2. Test-object luminescence intensity under the influence of a hydroxyapatite sample in a water medium with different exposition times.

We observed an increase in bio-luminescence both in the control and experimental samples when water medium was replaced with normal saline. Also, a decrease in the standard deviation value was registered which can be an indication of higher stability of the bacterial suspension (figure 3). The primary results descriptive analysis revealed no significant differences between the luminescence intensity of the test object and the control group, regardless of the exposition time (figure 3).

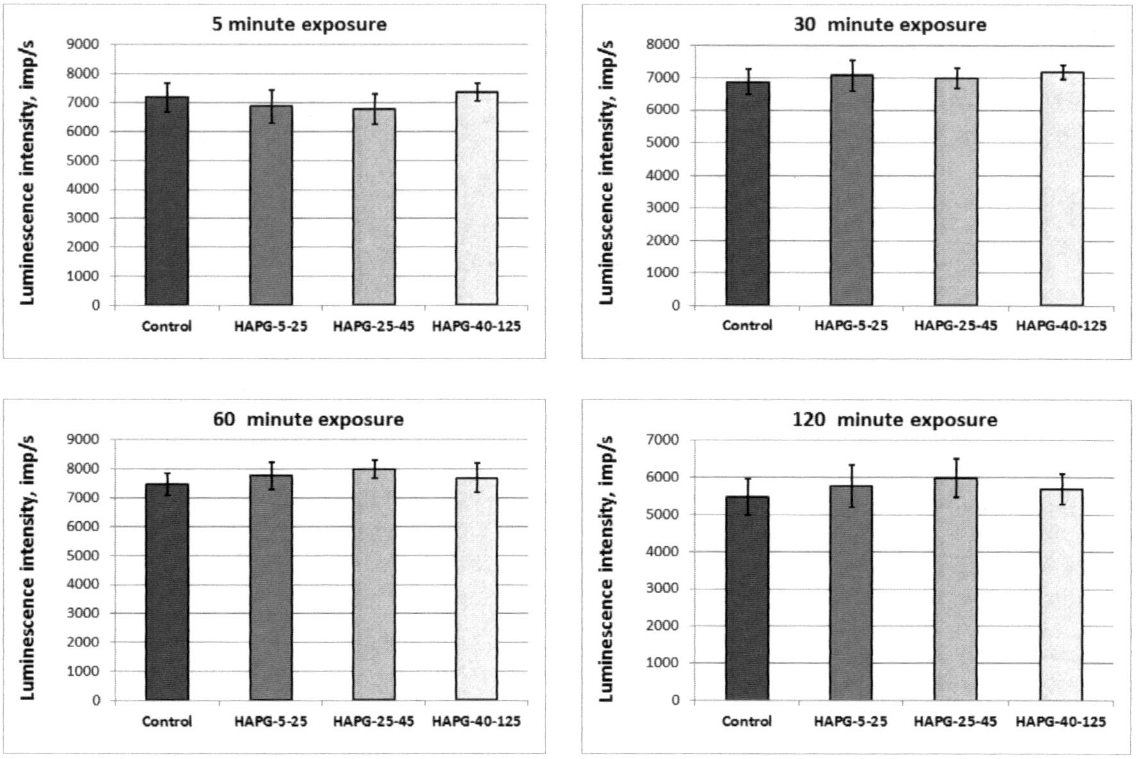

Figure 3. Luminescence intensity of the test subject treated with hydroxyapatite in sodium solution at different exposition times.

The data show that the hydroxyapatite samples produced no toxic effect on *E. coli* both in water and in normal saline, regardless of the period of exposition of the cells to the material and of the particle size distribution. However, toxic effect of hydroxyapatite in the same concentration on different test subjects was observed by a group of authors. In [11] genotoxic and cytotoxic effects as well asoxidative damage from increased concentrations of hydroxyapatite (5, 10, 20, 50, 75, 100, 150, 300, 500 and 1000 ppm) were studied in primary human blood cell cultures. In paper [12] toxic effect of nanosized hydroxyapatite at various concentrations (0 – 1000 μg cm($^{-2}$)) on a primary rat hepatocyte culture was evaluated. The author of [13] studied antimicrobial activity of hydroxyapatite nanoparticles against bacterial pathogens associated with implants. The mechanisms of nanoparticle effect on bacteria were studied by evaluating the generation of reactive oxygen species (ROS), DNA fragmentation, loss of lactate dehydrogenase (LDH), and cell interaction. Also, the cytotoxicity of hydroxyapatite particles was determined with the MTT test and fluorescence microscopic analysis. The difference in toxicity noted in these studies and in our work can be explained by the fact that we studied microgranular hydroxyapatite characterised by a lower cytotoxicity level than nanodispersed material examined by other research teams.

4. Conclusion

Thus, we have established that the studied hydroxyapatite samples with particle sizes of 5 – 25 μm, 25 – 45 μm and 40 – 125 μm have no toxic effect on human blood leukocytes (at a concentration of 40 mg per 1 ml of cell suspension), and on bacteria *E. coli* (at a concentration of 40 mg per 1 ml of water or normal saline), regardless of the time of contact of the cells with the material and the particle size distribution. The obtained data demonstrate safety of the examined materials and absence of toxic effects towards the test objects. This allows us to regard the studied hydroxyapatite samples as biocompatible.

Acknowledgments

The work was funded by the Ministry of Science and Higher Education (formerly, the Ministry of Education and Science) of the Russian Federation under Project No. RFMEFI57517X0168.

References

[1] Larry L H 1991 Bioceramics: from concept to clinic. *J Am Ceram Soc*, **74** 1487-1510
[2] Roveri N and Palazzo B 2007 Hydroxyapatite nanocrystals as bone tissue substitute *Nanotechnologies for the Life Sciences* **9** 283
[3] Lin K and Chang J 2015 Structure and properties of hydroxyapatite for biomedical applications *Hydroxyapatite (HAp) for biomedical applications* Woodhead Publishing p 3
[4] Okabayashi R, Nakamura M, Okabayashi T, Tanaka Y, Nagai A and Yamashita K 2009 Efficacy of polarized hydroxyapatite and silk fibroin composite dressing gel on epidermal recovery from full-thickness skin wounds *J Biomed Mater Res B* **90** 2 641
[5] Shin Y, Aoki H, Yoshiyama N, Akao M and Higashikata M 1992 Surface properties of hydroxyapatite ceramic as new percutaneous material in skin tissue *J Mater Sci-Mater M* **3** 3 219.
[6] Ji DY, Kuo TF, Wu HD, Yang JC and Lee SY 2012 A novel injectable chitosan/polyglutamate polyelectrolyte complex hydrogel with hydroxyapatite for soft-tissue augmentation *Carbohyd Polym* **89** 4 1123
[7] Liu M, Zhou, G, Song W, Li P, Liu H, Niu X and Fan Y 2012 Effect of nano-hydroxyapatite on the axonal guidance growth of rat cortical neurons *Nanoscale* **4** 10 3201
[8] Ko KS, Koh DC and Kong I 2017 Evaluation of the effects of nanoparticle mixtures on Brassica seed germination and bacterial bioluminescence activity based on the theory of probability *Nanomaterials* **7** 10 344
[9] Zakharova O V, Godymchuk A Y, Gusev A A, Gulchenko S I, Vasyukova I A and Kuznetsov D V 2015 Considerable variation of antibacterial activity of Cu nanoparticles suspensions depending on the storage time, dispersive medium, and particle sizes *Biomed Res Int* **2015** Article ID 412530
[10] Mogilnaya O A 2010 Growth and bioluminescence of luminous bacteria under the action of aflatoxin B1 before and after its treatment with nanodiamonds *Appl Biochem Micro+* **46** 1 33
[11] Turkez H, Yousef M I, Sönmez E, Togar B, Bakan F, Sozio P and Stefano AD 2014 Evaluation of cytotoxic, oxidative stress and genotoxic responses of hydroxyapatite nanoparticles on human blood cells *J Appl Toxicol* **34** 4 373
[12] Sonmez E, Cacciatore I, Bakan F, Turkez H, Mohtar Y I, Togar B and Stefano A D 2016 Toxicity assessment of hydroxyapatite nanoparticles in rat liver cell model in vitro *Hum Exp Toxicol* **35** 10 1073
[13] Baskar K, Anusuya T and Venkatasubbu G D 2017 Mechanistic investigation on microbial toxicity of nano hydroxyapatite on implant associated pathogens *Mat Sci Eng C-Mater* **73** 8

IOP Conf. Series: Materials Science and Engineering **731** (2020) 012021 doi:10.1088/1757-899X/731/1/012021

Nonwoven PTFE Membranes Fabricated by Electrospinning Method: Preparation and Characterization

I M Kolesnik[1], E N Bolbasov[1,2], V M Buznik[3] and S I Tverdokhlebov[1]

[1]National Research Tomsk Polytechnic University, 30, Lenin Ave., Tomsk, 634050, Russian Federation
[2]Microwave Photonics Lab, V.E. Zuev Institute of Atmospheric Optics SB RAS, 634055, Tomsk, Russian Federation
[3]All Russian Scientific Research Institute of Aviation Materials, 17, Radio Street, 105005, Moscow, Russian Federation

Ebolbasov@gmail.com

Abstract. Polytetrafluoroethylene (PTFE) is one of the promising materials for the purposes of tissue engineering and chemical technology because of its excellent physico-chemical properties and mechanical characteristics. However, conventional methods of PTFE porous membranes production have several disadvantages which limit the number of potential application areas by reason of a insufficient surface-to-volume ratio and poor porosity at small thickness. In the paper the results of using PTFE water suspension with a solution of water-soluble polymer for preparation of porous membranes by electrospinning are reported. The physico-chemical characteristics of membranes were investigated depending on the content of PTFE dispersion in spinning solution. There were found high hydrophobicity and lyophilicity of PTFE electrospun membranes. Main reasons of poor mechanical properties of porous membranes at high content of PTFE suspension in spinning solution are discovered. The ways of mechanical properties improvement and areas of possible applications are proposed.

1. Introduction

Materials with good mechanical characteristics and hydrophobic properties are widely used in chemical technology. Materials with similar features are applicable for the purification of gases from particles of organic and inorganic origin, osmotic distillation (OM), membrane distillation (MD), and their varieties. In the case of MD, important qualities are the filtration rate and high temperature gradient across the membrane thickness which can be achieved with high porosity and low thermal conductivity [1].

Polytetrafluoroethylene (PTFE) is a fluoropolymer synthesized by tetrafluoroethylene polymerization at high pressure [2]. Owing to its high chemical and thermal stability, good dielectric, hydrophobic properties, and excellent mechanical characteristics, PTFE is utilized in chemical technology for manufacturing of chemical reactors, chemical-resistant pipes, and membranes for various types of purification [3]. In reconstructive surgery and tissue engineering materials based on PTFE are used as a coating for blood vessels, bile ducts, larynx, implants, walls of heart valves, tissue engineering membranes in dentistry, plastic surgery [4].

The main methods for the manufacture of PTFE-based membranes are biaxial stretching, laser ablation, and the method using pore-forming reagents [5, 6]. The disadvantages of these methods are difficulties with manufacturing of thin membranes with a high degree of pore interconnectivity, PTFE depolymerization as a result of exposure to high-density laser radiation, usage of environmental-unfriendly plasticizers, and other reagents. Thus, the development of new methods for producing porous PTFE membranes is a necessary challenge.

The electrospinning of polymer solutions or melts is a promising method for production of porous polymer materials [7]. It provides capability to form constructions with high porosity, surface-to-volume ratio, and good mechanical characteristics for various applications: tissue engineering, regenerative medicine, drug delivery, biosensors, and membranes for MD, OD, and water purification [7]. Also, electrospinning makes possible to synthesize highly porous materials at small thickness that is needed for some membrane desalination methods. However, direct formation from PTFE solutions is impossible because of the lack of suitable solvents for PTFE and the high viscosity of its melt. The paper presents the results of the study on the use of PTFE suspensions with the solution of water-soluble fiber-forming polymer – polyvinyl alcohol (PVA) for the formation of membrane by electrospinning method.

2. Materials and methods

2.1. Preparation of PTFE-PVA precursor membranes by electrospinning

A spinning solution that was used to form nonwoven materials and comprised of a PTFE suspension F-4D (Halopolymer, Russia) and a 9 wt.% aqueous solution of polyvinyl alcohol (PVA) 16-1 (Evrohim, Russia). For investigation 4 groups of materials were formed in such way that the content of the PTFE suspension in the solution was 50, 60, 70, and 80 wt.%. The mixture was processed in an ultrasonic bath (Sapphire 5, Russia) for 10 hours at temperature of 30±3 °C until a homogeneous viscous liquid was obtained.

Nonwoven materials were formed by the electrospinning method on Nanon-01 (MECC, Japan) using a cylindrical collector with length – 200 mm and diameter – 100 mm under following technological parameters: voltage – 25 kV, flow rate – 2 ml/h, the needle – 22 G, the distance between needle and collector was 15 cm.

2.2. Sintering

After formation the precursor membranes were heated to 360±10 °C at a rate of 5±1 °C/h in atmospheric pressure, kept at a given temperature for 5 hours, and then cooled to room temperature at a speed of 5±1 °C/h.

2.3. Scanning electron microscopy (SEM)

SEM images allowed determine the average diameter of fibers and specific number of defects. Samples of each type of membranes were fixed onto metallic studs with double-sided conductive tape. A thin gold film was sprayed onto samples in order to provide a contact of the material with the stub and to prevent the accumulation of a negative charge on the samples surface. The morphology and chemical composition of the formed materials were studied using a scanning electron microscope (JEOL JCM-6000, Japan). The average fiber diameter and specific number of defects were calculated using ImageJ 1.5 software (National Institutes of Health, Bethesda, MD, USA).

2.4. Porosity

The porosity was determined by measuring the geometric dimensions and weight of the membranes after sintering based on which the calculation was carried out according to the formula:

$$P=\left(1-\frac{\rho_{exp}}{\rho}\right)\cdot 100\ \%,$$

where ρ_{exp} – calculated mean of density, ρ – density of PTFE.

2.5. Mechanical properties
The investigation of strength and elongation under uniaxial tension was studied according to GOST R 53226-2008 using a tensile testing machine (Instron 3344, USA).

2.6. Surface wetting
Surface wetting with water and T-1500 transformer oil (GOST 982-80) was investigated using an optical goniometer (Easy Drop-100, Germany).

2.7. FTIR spectroscopy
Chemical structure of the fabricated porous materials before and after sintering was investigated by Fourier transform infrared (FTIR) spectroscopy using Cary 630 spectrometer (Agilent, USA) in the range of 500 to 4000 cm^{-1} with a resolution of 4 cm^{-1}.

3. Results and discussion
Images of the nonwoven materials surface depending on the weight content of the PTFE suspension in the spinning solution before (left column) and after heat treatment (right column) are shown in figure 1. The initial polymer membranes are formed by randomly interwoven fibers with regular cylindrical shape. On the surface of the fibers there are no defects, such as breaks and cracks, that indicates the optimally selected technological parameters of membrane formation. The interwoven fibers form interconnected pores that is typical for nonwoven materials fabricated by electrospinning [8].While the content of the PTFE suspension in the spinning solution increases, the average fiber diameter decreases because the solution viscosity lowers (table 1).

Heat treatment of samples at temperature of 360 °C leads to changes in the morphology of the fibers that form the polymer membrane (figure 1). Thickening and cracking occur on the fibers surface. The number of these changes per unit length of fiber increases with raising content of the PTFE suspension in the solution (table 1). The porosity of the obtained membranes also does not change depending on the content of the suspension and is equal to 83±2%.

Table 1. Physico-chemical properties of PTFE porous materials depending on PTFE suspension content in spinning solutions.

Content of PTFE suspension wt.%	Viscosity mPa·s	Average fiber diameter µm		Specific number of defects, μm^{-1}	Mechanical properties	
		before sintering	after sintering		Tensile strength MPa	Elongation %
50	414±12	0.96±0.18	1.02±0.18	0.45±0.06	0.45±0.02	115±16
60	332±10	0.94±0.14	0.90±0.17	0.64±0.09	0.31±0.04	113±8
70	217±6	0.89±0.16	0.95±0.19	0.71±0.06	0.13±0.01	63±5
80	136±4	0.77±0.16	0.96±0.18	0.93±0.08	0.06±0.01	43±4

The formation of defects in the fiber is explained by the following main reasons. The first is the decomposition of PVA, surfactants, and the removal of water from the fiber bulk during heat treatment. The removal of volatile impurities leads to the formation of voids between the particles of PTFE. The filling of these cavities is obstructed even at elevated temperatures in consequence of the high PTFE melt viscosity. Secondly, PTFE is a semi-crystalline polymer which crystallinity during crystallization from the melt is determined by the cooling rate, while the maximum crystallization rate is observed at 310-315 °C. Considering that the cooling rate of the samples after sintering was 5±1 °C/h, favorable conditions were selected during the cooling process for the formation of the crystalline phase in PTFE. Since the crystalline phase occupies a smaller volume than the amorphous phase, significant stresses appear in the polymer that stimulate the defects formation in polymer fibers. At the same time, formed during the destruction process the low molecular weight products of PVA and surfactants can diffuse into the structure of PTFE and have a plasticizing effect thereby complicating the crystallization process. Thus, fewer defects and, as a result, better physico-mechanical properties are observed for membranes obtained from spinning solutions with a low concentration of PTFE suspension, since the amount of plasticizer in these samples is greater (table 1).

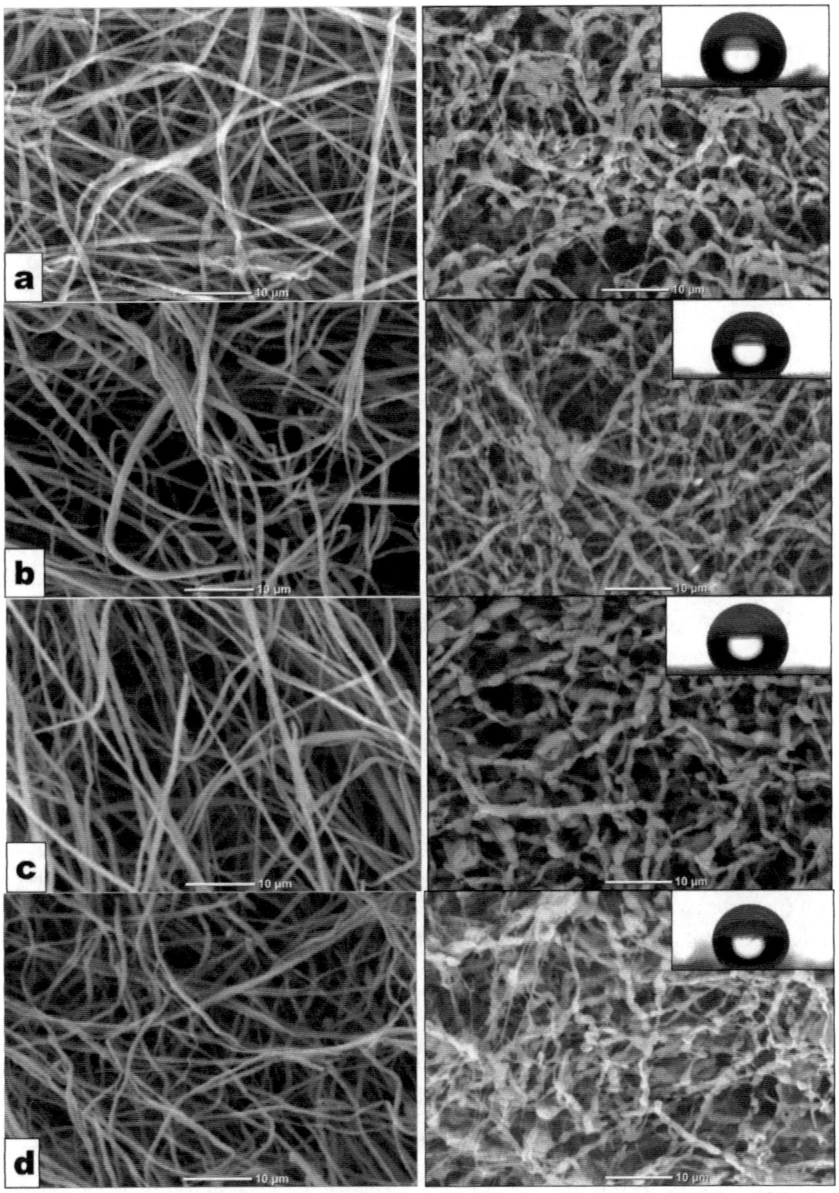

Figure 1. SEM images of nonwoven materials formed from spinning solutions with a suspension of PTFE from a) 50 to d) 80 wt. % (left column – precursor materials, right column – after sintering).

A study of the surface wetting of the formed porous materials indicates that, regardless of the suspension content, all formed membranes are hydrophobic, while the value of the contact angle of the surface for wetting with water is ~ 125° and does not significantly change when the content of the PTFE suspension in the spinning solution is varied (figure 1). Before sintering membranes completely absorb water droplets placed on the surface. In this case, a drop of transformer oil deposited on the membranes surface, regardless of the content of the PTFE suspension, is completely absorbed into the membrane after 2-3 seconds that indicates the high lyophilic properties of the formed membranes.

Confirmation of the removal of water and PVA from PTFE membranes after sintering was approved by FTIR spectroscopy, results of which are presented in figure 2. In the FTIR spectrum of PTFE precursor membranes (figure 2a) the highest absorbance is observed at 1203 and 1147 cm^{-1} that corresponds to asymmetric and symmetric stretching of $C–F_2$ in PTFE [9]. At 3325, 2912, 1457, and 838 cm^{-1} the stretching of O–H, asymmetric stretching of $C–H_2$, asymmetric bending of $C–H_2$, and stretching of C–C, respectively, are detected that correlate with the presence of PVA in the precursor membranes [10, 11]. After sintering process (figure 2b) the strong absorbance peaks at 1204 and 1147 cm^{-1} are only observed that approve full decomposition of PVA.

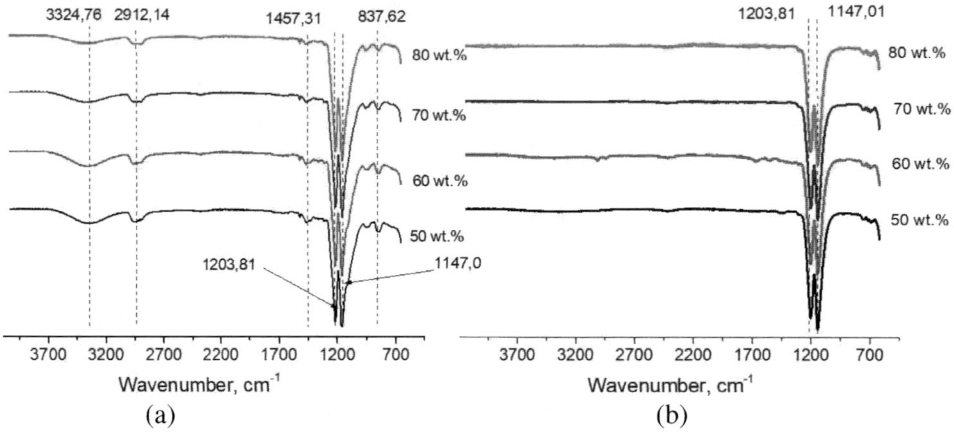

Figure 2. FTIR spectra of PTFE porous materials: (a) – spectrum of precursor materials, (b) – spectrum of materials after sintering.

Conclusion

In conclusion, porous materials based on PTFE were fabricated by electrospinning with sintering of precursor membranes. It was demonstrated that the obtained materials have high porosity with pore interconnectivity, as well as good mechanical characteristics. High hydrophobicity and lipophilicity of the obtained membranes because of the low surface energy of PTFE and high porosity of the membrane were found [12]. After sintering process the full decomposition of PVA and surfactant was observed. The combination of high hydrophobicity and porosity allows to utilize obtained membranes as selective ones, for example, for cleaning oils from water contaminants.

Future outcome

The improvement of the mechanical properties of PTFE membranes is associated with a decrease in the crystallinity of the membranes because of the optimization of heat treatment and the synthesis of composite membranes based on PTFE and carbon nanotubes.

Acknowledgments

This work was financially supported by the Ministry of Science and Higher Education of the Russian Federation, Federal Target Program (agreement # 14.575.21.0140, unique identifier RFMEFI57517X0140).

References

[1] Huang Y, Huang Q-L, Liu H, Zhang C-X, You Y-W, Li N-N and Xiao C-F 2017 *Journal of Membrane Science* **523** 317

[2] Cui Z, Drioli E and Lee Y M 2014 *Prog. Polym. Sci.* **39** 164
[3] Feng S, Zhong Z, Wang Y, Xing W and Drioli E 2018 *Journal of Membrane Science* **549** 332
[4] Carbonell J M, Martín I S, Santos A, Pujol A, Sanz-Moliner J D and Nart J 2014 *Int. J. Oral Maxillofac. Surg.* **43** 75
[5] Guo Y, Chen J, Hao X, Zhang J, Feng X and Zhang H 2007 *J. Mater. Sci.* **42** 2081
[6] Grakovich P N, Ivanov L F, Kalinin L A, Ryabchenko I L, Tolstopyatov E M and Krasovskii A M 2009 *Russ. J. Gen. Chem.* **79** 626
[7] Ahmed F E, Lalia B S and Hashaikeh R 2015 *Desalination* **356** 15
[8] Bolbasov E N, Stankevich K S, Sudarev E A, Bouznik V M, Kudryavtseva V L, Antonova L V, Matveeva VG, Anissimov Y G and Tverdokhlebov S I 2016 *Mater. Chem. Phys.* **182** 338
[9] Wang C and Chen J-R 2007 *Applied Surface Science* **253** 4599
[10] Li J-H, Yan B-F, Shao X-S, Wang S-S, Tian H-Y and Zhang Q-Q 2015 *Applied Surface Science* **324** 82
[11] Krimm S, Liang C Y and Sutherland G B B M 1956 *Journal of Polymer Science* **22** 227
[12] Cassie A B D and Baxter S 1944 *Trans. Faraday Soc.* **40** 546

Composition Polymeric Membranes Based on the VDF-TeFE Copolymer Formed by Electrospinning

T S Tverdokhlebova[1], E N Bolbasov[1] and V M Bouznik[2]

[1]Tomsk Polytechnic University, Lenin Ave., 30, 634050, Tomsk, Russia
[2]All Russian Scientific Research Institute of Aviation Materials, Budenny Ave., 25a, Building 3, Moscow

aramat_tts@mail.ru, ebolbasov@gmail.com

Abstract. The paper presents the results of pilot studies on the formation of non-woven hydrophilic piezoelectric composite polymer membranes based on a copolymer of vinylidene fluoride and tetrafluoroethylene (VDF-TeFE) and polyethenylpyrrolidine (PEP) by electrospinning. Using optical goniometry, the effect of the PEP content on the wetting of the membrane surface with water was established. Using scanning electron microscopy, the influence of the content PEP on the structure of the formed membranes was studied. The effect of an isotonic NaCl solution on the structure of the formed membranes upon their dissolution within 7 days was studied. Based on studies of the chemical composition of the membranes before and after their dissolution, an assumption was made about the possibility of using the developed membranes for targeted delivery of pharmacological preparations for wound healing.

1. Introduction

The first wounds and wound treatments were described 5,000 years ago. Progress that goes beyond the ancient practice of wound care is a recent phenomenon unlike a large number of common technological inventions over the past 100 years. Chronic wound healing is a complex laborious process that is susceptible to external biological attacks (healing time ~ 12 weeks), therefore it is necessary to create a controlled environment for wound healing using an appropriate "barrier".

Promising materials for creating a "barrier" are nonwoven polymer membranes formed by the electrospinning method [1]. It is known that piezoelectric membranes made of fluorine-containing polymers, under the influence of external mechanical stimuli, can enhance migration, adhesion, and the level of secretion of fibroblast cytokines, the main skin cells [2]. This ability allows piezoelectric fluoropolymer membranes (FPMs) to generate electrical impulses as a result of mechanical action on the tissues surrounding the implant, which allows to stimulate the healing process [3].

The drawback of the FPM, limiting their use as wound dressings, is the high hydrophobicity, significantly reducing the ability of the FPM to absorb exudate (a fluid rich in protein and cellular elements that oozes out of blood vessels due to inflammation and is deposited in nearby tissues). One of the possible solutions to the problem of creating effective FPMs for wound healing is the development of composite membranes based on fluorine and hydrophilic polymers.

In this article, we present the results of pilot studies on the creation by electrospinning of hydrophilic composite membranes based on a piezoelectric fluorine-containing polymer. Studies of the effect of the content of a hydrophilic component in a composite material on the structure, chemical composition and surface properties of membranes. The results of studies of the dissolution of formed membranes in simulating body fluid are also presented.

2. Materials and methods

A vinylidene fluoride-tetrafluoroethylene copolymer (VDF-TFE) (Mn 5×10^5, TFE 7 mol %) and polyethenylpyrrolidine (PEP) (Mn 3×10^3) were used to make the FPMs. A mixture of dimethylformamide (C_3H_7NO), dimethyl sulfoxide (C_2H_6OS) and acetone (C_3H_6O), taken in a ratio of 40/40/20 mass %, was used as a mixed solvent of polymers for the preparation of spinning solutions. Spinning solutions containing polymers taken in a ratio of 100; 95/5; 85/15; 75/25; 50/50 mass %, respectively, with a total polymer content in the dope of 7 mass %, were prepared in a sealed glass reactor at a temperature of 40°C and stirring continuously until a uniform, clear, viscous liquid was obtained. A commercially available NANON 01A installation (MECC Co., Japan), equipped with a cylindrical assembly collector with a diameter of 50 mm and a length of 200 mm, was used for the formation of nonwoven materials. The distance between the dope injector (22G needle) and the assembly manifold was 150 mm. The voltage at the injector is 25 kW. The flow rate of the dope solution is 2 ml/hour. Assembly speed 50 rpm.

To study the effect of simulating body fluid on the structure of the formed membranes, membranes 30×30 mm in size were prepared. The resulting membranes were filled with 40 ml of isotonic sodium chloride solution. Samples were kept in solution for 7 days at a temperature of 37±1 °C.

The structure of the formed membranes was studied by scanning electron microscopy (JEOL JCM-6000, Japan). To improve image quality, a thin layer of gold was applied to the sample surface for 2 minutes before examination using a SC7640 magnetron sputtering system (Quorum Technologies Ltd, England). The fiber diameter was determined from the SEM images out of 3 different fields of view using plugin DiameterJ, which located in Image J 1.38 (National Institutes of Health, USA) software.

EDS (JEOL JED 2300, Japan) carried out the study of the chemical composition of the membranes. Semi-quantitative chemical composition was assessed using the method of three corrections (ZAF-corrections): in the average atomic number, absorption and fluorescence.

The wetting of the surface of the materials with water until dissolution was investigated by the “sedentary drop” method using an optical goniometer (Easy Drop-100, Germany). For this, a drop of deionized water with a volume of 3 μl was placed on the membrane surface. The measurement of the contact angle was carried out three minutes after the contact of the liquid and the membrane.

3. Experiment results

It was revealed that membranes prior to dissolution with a PEP content in the range from 0 to 25 mass % are formed by interwoven fibers with a normal diameter distribution, have well-developed interconnected porosity (Figure 1, Table 1), while an increase in PEP content leads to a decrease the average diameter of the fibers (d_{av}) forming the membrane by more than 20% (Table 1). The decrease in d_{av} of the fibers forming the membrane is due to a decrease in the viscosity of the dope with an increase in the content of PEP. An increase in the PEP content to 50 mass % causes significant changes in the structure of the fibers forming the membrane, the fibers are grouped into bundles, while the bundles have a pronounced directivity (Figure 1 e) and d_{av} decreases by more than 25% from the level of control values. Probably, the reason for the change in the membrane structure in this case, in addition to reducing the viscosity of the dope solutions, is a significant change in the electrophysical characteristics (dielectric constant, conductivity, dipole moment) of the dope solution.

IOP Conf. Series: Materials Science and Engineering **731** (2020) 012022 doi:10.1088/1757-899X/731/1/012022

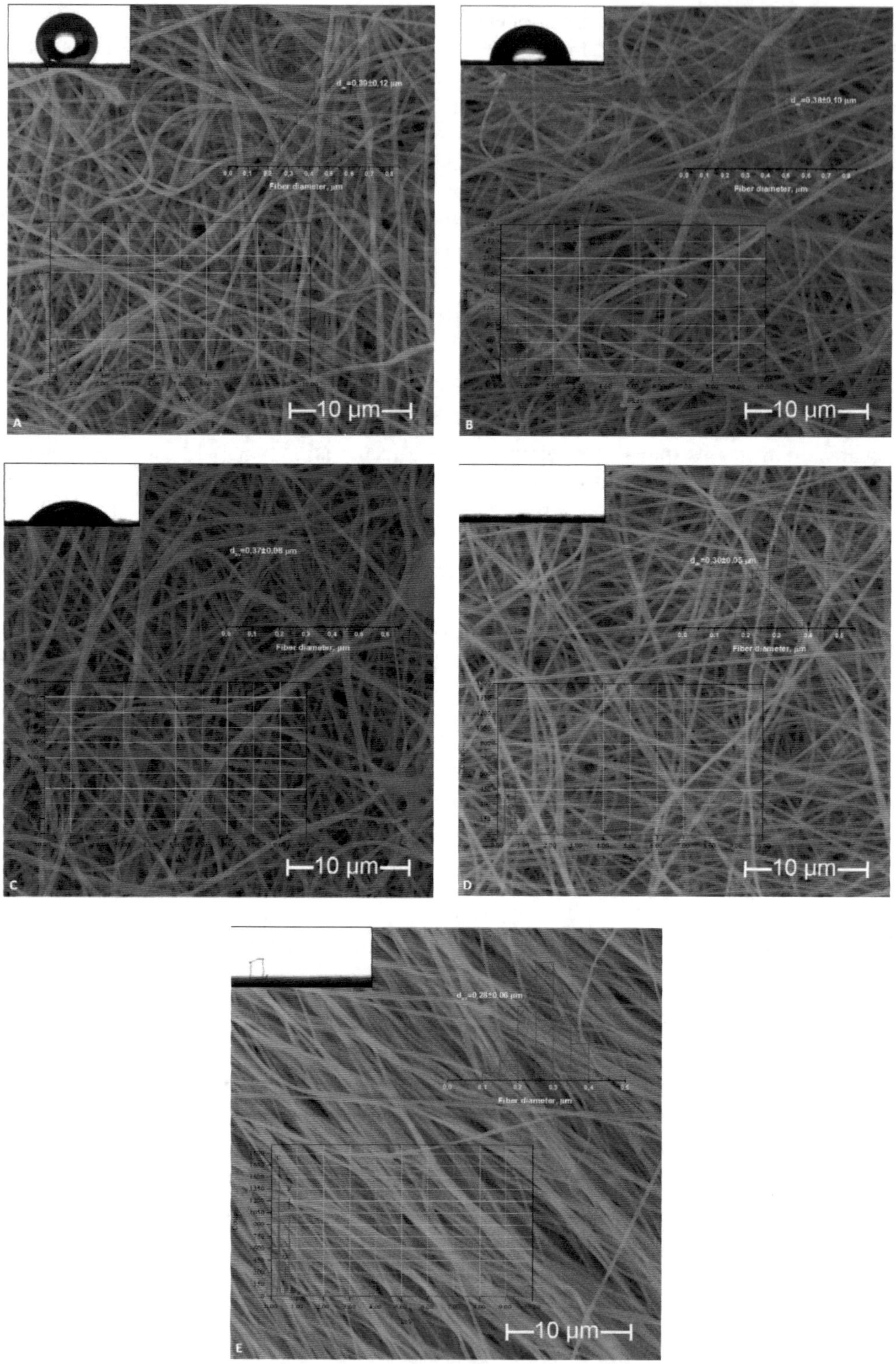

Figure 1. SEM images of samples: a) pure VDF-TeFE; b) VDF-TeFE containing 5% PEP; c) VDF-TeFE containing 15% PEP; d) VDF-TeFE containing 25% PEP; e) VDF-TeFE containing 50% PEP before dissolving.

IOP Conf. Series: Materials Science and Engineering **731** (2020) 012022 doi:10.1088/1757-899X/731/1/012022

An increase in the PEP content in the spinning solution leads to changes in the chemical composition of the membranes, increasing the oxygen content by a factor of ~ 21 compared to the control membrane. In this case, the contact angle of the membrane surface with water changes from 125 (super-hydrophobic surface) to 0 (full wetting, absorption of a drop of water into the membrane), which is associated with an increase in oxygen-containing groups on the membrane surface.

Table 1. Physical properties of samples.

PEP content (%)	Average diameter before dissolving (μm)	Average diameter after dissolving (μm)	Contact angle before dissolving (deg)
0	0.39±0.12	0.37±0.08	125±7
5	0.38±0.10	0.36±0.09	46±5
15	0.37±0.08	1.07±0.56	20±4
25	0.30±0.05	0.97±0.42	0
50	0.28±0.06	2.17±1.32	0

Exposure of membranes in an isotonic NaCl solution for 7 days causes structural changes in the formed membranes depending on the content of the hydrophilic component. Thus, the effect of an isotonic solution does not cause structural changes and significant changes in the average fiber diameter in membranes based on pure VDF-TeFE copolymer and in membranes with a PEP content of 5 mass% (Figure 1 a, b, Figure 2 a, b). With an increase in the PEP content in the composite membrane to 15%, the diameter of the fibers forming the membrane increases by more than 2.5 times compared with the membrane before immersion in an isotonic solution (Figure 1, Figure 2, Table 1). In this case, the fibers begin to "stick together" at the contact points (Figure 2 c). At a PEP content of 25 mass %, the average fiber diameter after dissolution increases by more than 3 times, while the membrane morphology is similar to membranes with a PEP content of 15 mass % (Figure 1 e, Figure 2 e, c). The greatest changes in the structure of the fibers forming the membrane are observed at a PEP content of 50 mass %. In this case, the average fiber diameter increases by more than 7.5 times. If, before immersion in an isotonic solution, the composite membrane was formed by separate directional fibers of the fiber, then after exposure to the solution the fibers stick together not only at the contact points, but also along the length of the fiber, which leads to a loss of fiber orientation and the formation of enlarged fibrous structures.

Table 2. Chemical composition of samples.

Before dissolving

PEP content (%)	Elements (mass %) C	O	F	C/F
0	43.86±0.62	0.41±0.25	55.67±0.51	0.79
5	43.60±1.58	1.90±1.04	54.50±1.14	0.80
15	46.02±1.07	3.03±0.06	50.95±1.02	0.90
25	49.18±0.37	4.08±0.82	46.74±0.61	1.05
50	52.90±0.44	8.64±1.76	38.45±2.12	1.38

After dissolving

PEP content (%)	Elements (mass %) C	O	F	Na	Cl	C/F
0	41.85±1.50	0.52±0.21	56.59±0.76	0.64±0.67	0.39±0.51	0.74
5	43.01±1.21	2.78±1.29	53.99±1.85	0.18±0.12	0.04±0.04	0.80
15	45.23±0.58	2.80±0.08	50.49±0.56	0.62±0.20	0.85±0.53	0.90
25	46.05±0.79	3.89±0.59	49.32±0.27	0.25±0.17	0.27±0.21	0.93
50	49.43±1.10	6.43±1.05	42.76±0.09	0.67±0.02	0.72±0.13	1.16

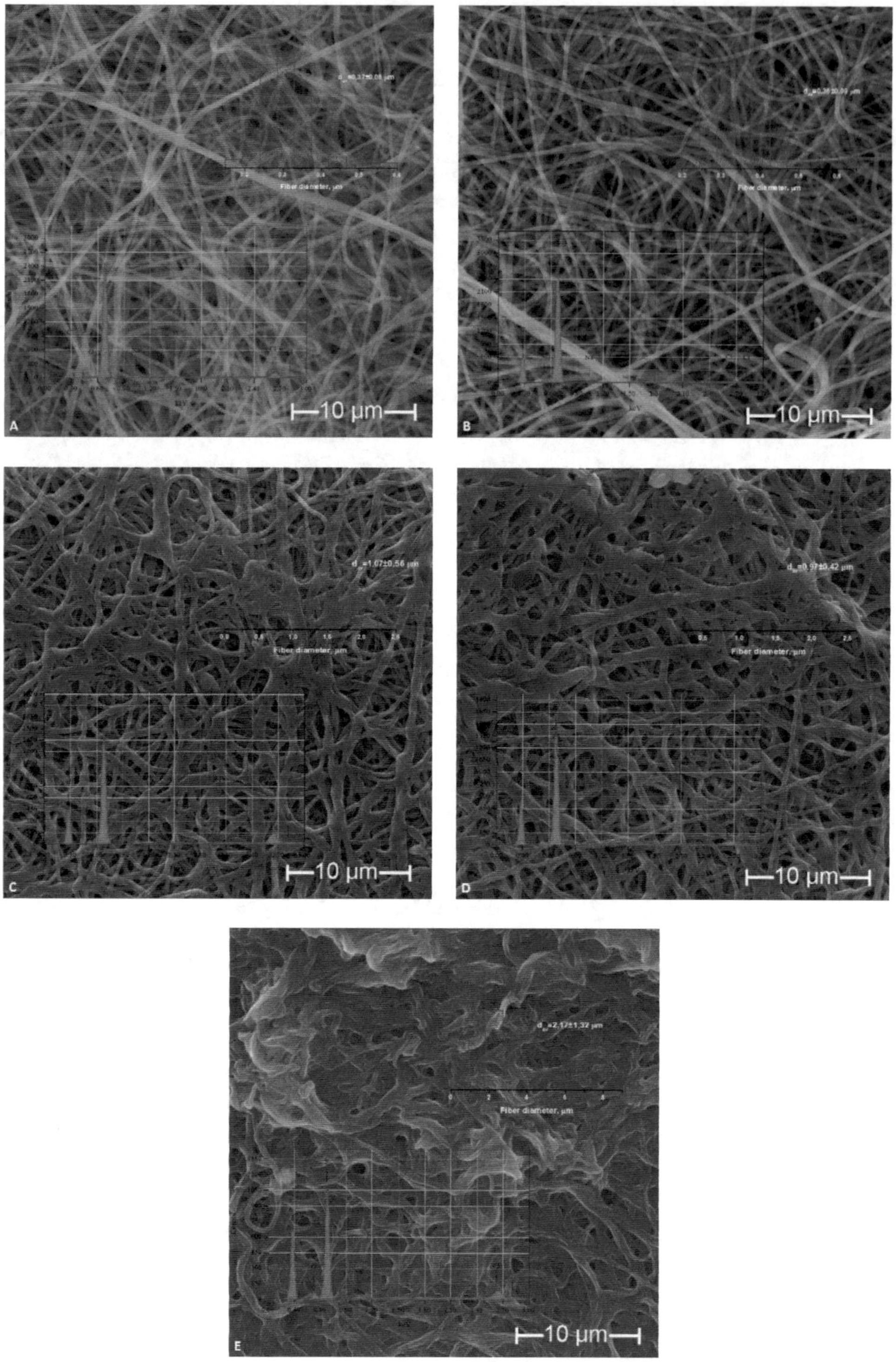

Figure 2. SEM images of samples: a) pure VDF-TeFE; b) VDF-TeFE containing 5% PEP; c) VDF-TeFE containing 15% PEP; d) VDF-TeFE containing 25% PEP; e) VDF-TeFE containing 50% PEP after dissolving.

The interaction of membranes with isotonic NaCL solution also causes changes in the chemical composition of the membranes (Table 2). Thus, membranes formed from pure VDF-TeFE copolymer retain their chemical composition, which is due to the high hydrophobicity and chemical resistance of the VDF-TeFE copolymer, which prevents the dissolution of the membrane. The presence in the membranes based on the VDF-TeFE copolymer of elements of an isotonic Na and Cl solution is due to the crystallization of these elements on the membrane surface, since the membrane floated on the solution surface during the entire interaction time. When the PEP content in the composite membrane is 5 mass %, no significant changes in the chemical composition of the membranes after interaction with the solution are also observed. This is due to the fact that the concentration of the hydrophilic component is not significant and the hydrophilic component is evenly distributed in the matrix of the hydrophobic component of the composite. Under these conditions, the interaction of water and PEP is possible only on the surface of the fibers forming the membrane; therefore, the output of PEP into the solution is significantly limited. When the PEP content in the composite is 15 mass %, the concentration of PEP on the surface of the fibers increases, however, most of the molecules are still firmly bound to the polymer matrix, which complicates their migration to the solution. At the same time, part of the "free" polymer chain of PEP is capable of sorbing water from the solution, which leads to an increase in fibers diameter with a constant C/F ratio. When the PEP concentration in the composite is 25 mass %, conditions arise when the PEP molecules are able to leave the fibers, which leads to a decrease in the C/F ratio, while the concentration of PEP molecules capable of absorbing water from the solution increases both on the surface and in the volume of the fibers, which is also leads to an increase in the diameter of the fibers forming the membrane. The greatest difference in C/F ratios before and after interaction with an isotonic solution is observed in composites with a PEP content of 50 mass %, which is due to the largest number of PEP molecules capable of leaving the polymer fibers. Also in these composites there is the largest number of PEP molecules capable of absorbing water, which also explains the largest changes in the morphology of these membranes.

4. Conclusion

The possibility of creating composite polymer hydrophilic membranes based on a copolymer of vinylidene fluoride with tetrafluoroethylene and polyethenylpyrrolidine (PEP) by the electrospinning method has been shown for the first time. It was established that an increase in the concentration of PEP in the composite material allows one to change the contact angle of the membranes with water from 125 to 0 degrees. Using scanning electron microscopy, it was found that with an increase in the concentration of PEP, the fibers forming the membrane decrease from 0.39 ± 0.12 to 0.28 ± 0.06 µm. It was found that significant changes in the average diameter of the fibers forming the membrane during the interaction of the membrane with an isotonic NaCl solution are observed at a PEP concentration of 15 mass %. The study of the chemical composition of the membranes by EDX revealed a decrease in the C/F ratio after dissolution, which suggests the use of the developed membranes for targeted delivery of pharmacological preparations.

Acknowledgments
The work performed according to the project of fundamental research of the Russian Academy of Sciences № 0293-2019-0006.

References

[1] Ambekar R S and Kandasubramanian B 2019 Advancements in nanofibers for wound dressing: A review *Eur. Polym. J.* **117** 304–36
[2] Guo H-F, Li Z-S, Dong S-W, Chen W-J, Deng L, Wang Y-F and Ying D-J 2012 Piezoelectric PU/PVDF electrospun scaffolds for wound healing applications *Colloids Surfaces B Biointerfaces* **96** 29–36
[3] Wang A, Liu Z, Hu M, Wang C, Zhang X, Shi B, Fan Y, Cui Y, Li Z and Ren K 2018 Piezoelectric nanofibrous scaffolds as in vivo energy harvesters for modifying fibroblast alignment and proliferation in wound healing *Nano Energy* **43** 63–71

IOP Conf. Series: Materials Science and Engineering **731** (2020) 012023 doi:10.1088/1757-899X/731/1/012023

The pH Level Influence on Hydroxyapatite Phase Composition Synthesized with Hydrothermal Method

K Chuprunov[1], A Yudin[1], D Lysov[1], E Kolesnikov[1], D Kuznetsov[1], D Leybo[1], I Ilinykh[1] and A Godymchuk[2,3]

[1]Department of Functional Nanosystems and High-Temperature Materials, National University of Science & Technology "MISIS", 4 Leninsky avenue, Moscow 119049, Russia
[2]Materials Science Department, Tomsk Polytechnic University, 30 Lenina street, Tomsk 634050, Russia
[3]Chemistry Environmental Laboratory, Tobolsk Complex Scientific Station, Ural Branch of the Russian Academy of Science, 15 Akademician Osipov Street, Tobolsk 626152, Russia

E-mail: yudin@misis.ru

Abstract. This paper reports the pH level influence on hydroxyapatite phase composition synthesized with hydrothermal method in $Ca(OH)_2$-H_3PO_4, $Ca(NO_3)_2$-$(NH_4)_2HPO_4$-NH_4OH, $Ca(OH)_2$-$NH_4H_2PO_4$.The obtained samples were studied with X-Ray diffraction, Fourier Transform Infrared (FTIR) and Raman spectroscopy. The one phase $Ca_5H_2O_{13}P_3$ high crystallinity hydroxyapatite was synthesized with hydrothermal method at pH equal to 11. The crystallinity degree was calculated from the X-Ray diffraction pattern and became 0.96. The increasing pH level from 7 to 11 provides obtaining one phase hydroxyapatite at pH level 11 instead the two phase $Ca_{9.04}(PO_4)_6(OH)_{1.68}$, $CaHPO_4$ at pH level 9 and $CaPO_3(OH)$, $Ca(OH)_2$ at pH level 7.

1. Introduction

Biomaterials attract increasing interest owing to their applicability in a field of medicine [1]. The hydroxyapatite materials due to specific properties [2] find application as a biomedical material [3]. The hydroxyapatite properties, like crystallinity degree, crystal dispersity, porosity and stoichiometry make hydroxyapatite [4] widely used as scaffold for tissue engineering [4], fillers for healing defect in bones [5] and drug delivery systems [6].

Ceramic materials are a class of biomaterials used in biomedical devices. Ceramics based materials are widely used as implant materials due to their ability to be fabricated into a variety of shapes, along with their variable porosity, bioactive properties in the body and high compressive strength [7]. The calcium phosphates and hydroxyapatite have close chemical composition to human bone minerals [8]. Ceramic materials based on calcium phosphate and hydroxyapatite demonstrate high biocompatibility, perfect osteoconduction and bioactivity [9]. Also HAp finds application as biocompatible cover for implants made of metal materials [10]. The hydroxyapatite $Ca_{10}(PO_4)_6(OH)_2$ is the most thermodynamically stable in its crystalline state. HAp could be integrated with bones without any local toxicity and inflammation of body response [11].

The synthesis methods and synthesis parameters provide the phase composition, morphological and dispersity control of the obtained materials. The two ways of hydroxyapatite fabrication exist extracting from nature sources and synthetic by different methods dry, wet and high temperature methods [9]. The HAp could be synthesized by the various methods including solid-state method [12], mechanochemical [13], chemical precipitation [14], hydrolysis [15], sol-gel [16], hydrothermal [17], emulsion [18], sonochemical [19], combustion [20], pyrolysis [21], synthesis from biogenic sources [22] and combined methods [23] methods. Any method provides its own special phase composition, dispersion and morphology of the hydroxyapatite. The hydroxyapatite properties depend on the synthesis parameters.

The aim of the presented paper is to estimate the initial precursor influence in precursor systems $Ca(OH)_2$-H_3PO_4, $Ca(NO_3)_2$-$(NH_4)_2HPO_4$-NH_4OH and $Ca(OH)_2$-$NH_4H_2PO_4$ on the hydroxyapatite phase composition, dispersion and crystallinity degree. Also, the pH level influence during synthesis process described at [24, 25] on the hydroxyapatite samples properties.

This paper discuss the first time using of chemical precipitation and hydrothermal synthesis method to obtain the calcium phosphate and hydroxyapatite powders. The synthesized hydroxyapatite powder will be used as an initial material in granulation process to produce hydroxyapatite spherical granules for 3D printing of the ceramic implants.

2. Materials and Methods

2.1. Hydrothermal synthesis of hydroxyapatite sampleshe

Calcium nitrate tetrahydrate ($Ca(NO_3)_2{\bullet}4H_2O$, AR Grade), diammonium hydrogen phosphate $(NH_4)_2HPO_4$, AR Grade) ammonium hydroxide (NH_4OH, AR Grade), calcium hydroxide ($Ca(OH)_2$, AR Grade), phosphoric acid (H_3PO_4 AR Grade) produced by Reachem Co. (Russia) were used as initial precursors for obtaining HAp samples. The HAp samples were synthesized in three systems: $Ca(OH)_2$-H_3PO_4, $Ca(OH)_2$-$NH_4H_2PO_4$ and $Ca(NO_3)_2$-$(NH_4)_2HPO_4$-NH_4OH. In $Ca(OH)_2$-H_3PO_4 system calcium hydroxide was added to the phosphoric acid. The Ca/P element ratio was calculated equal to 1.67. Calcium hydroxide was added to diammonium hydrogen phosphate in system $Ca(OH)_2$-$NH_4H_2PO_4$. In $Ca(NO_3)_2$-$(NH_4)_2HPO_4$-NH_4OH system the diammonium hydrogen phosphate solution was added to the calcium nitrate solution and ammonium hydroxide was used to increase pH level. The obtained suspensions were treated in stainless steel autoclave at 250 °C with 150 bar during 24 hours. The obtained suspensions were washed with distilled water till pH level became equal to 7 by decantation method.

The next step was samples washing suspensions with decantation method until pH level became equal to 7 using the distilled water. The vacuum filtration system separated the HAp samples from the dispersant liquid. The HAp samples dried at 90 °C during 12 hours in the drying chamber SNOL HS-80-01. The obtained samples are presented in table 1.

Table 1. The obtained samples of hydroxyapatites.

Sample	Synthesis condition	
	System	pH
HAp 1.1	$Ca(OH)_2$-H_3PO_4	7
HAp 1.2	$Ca(OH)_2$-$NH_4H_2PO_4$	7
HAp 1.3	$Ca(NO_3)_2$-$(NH_4)_2HPO_4$-NH_4OH	7
HAp 1.4	$Ca(NO_3)_2$-$(NH_4)_2HPO_4$-NH_4OH	9
HAp 1.5	$Ca(NO_3)_2$-$(NH_4)_2HPO_4$-NH_4OH	11

2.2. Analytical methods to study hydroxyapatite samples

Phase composition and crystallinity degree were characterized with X-Ray diffractometer Scientific Instruments Difray 401 (Russia) with Cr Kα emission (wave length of 2.2909 Å) using PDF-2004 and

IOP Conf. Series: Materials Science and Engineering **731** (2020) 012023 doi:10.1088/1757-899X/731/1/012023

COD (Crystallography Open Database). The diffraction patterns were detected in the range from 14 to 140° 2θ with the interval 58 degrees and equilibrium time 300 second. The crystallinity degree (X_c) was calculated with the equation (1). The $V_{112/300}$ meant the minimum intensity between two intensity peaks reflected from the crystallography planes (112) and (300) consequently. The I_{300} meant the maximum intensity for the peak reflected from crystallography plane (300).

$$X_c = 1 - \frac{V_{112/300}}{I_{300}}$$

Also, the Fourier Transform Infrared spectroscopy (FTIR spectroscopy) Thermo Scientific Nicolet 380 (USA) and Raman spectroscopy Thermo Scientific DXR Microscope (USA) with OMNIC operation system were used to determine hydroxyapatite samples phase composition. The specters determination was carried out via book [26].

3. Results and discussion

The Figure 1 demonstrates XRD patterns of the hydroxyapatite samples obtained via hydrothermal method. The HAp 1.1 sample is one phase $Ca_3(PO_4)_2$ [00-009-0169]. In the sample HAp 1.2 $CaHPO_4$ [00-001-0653] and hydroxyapatite $Ca_{10}(PO_4)_6(OH)_2$ [01-72-1243] phases were detected. Also, the HAp 1.3 sample consists from two phases $Ca(OH)_2$ [00-050-0008] and $CaPO_3(OH)$ [00-009-0080]. The HAp 1.4 sample has two phase composition with $CaHPO_4$ [00-003-0423] and hydroxyapatite phase $Ca_{9.04}(PO_4)_6(OH)_{1.68}$ [01-086-1203]. The sample HAp 1.5 is one phase sample hydroxyapatite $Ca_5H_2O_{13}P_3$ [96-900-2215].

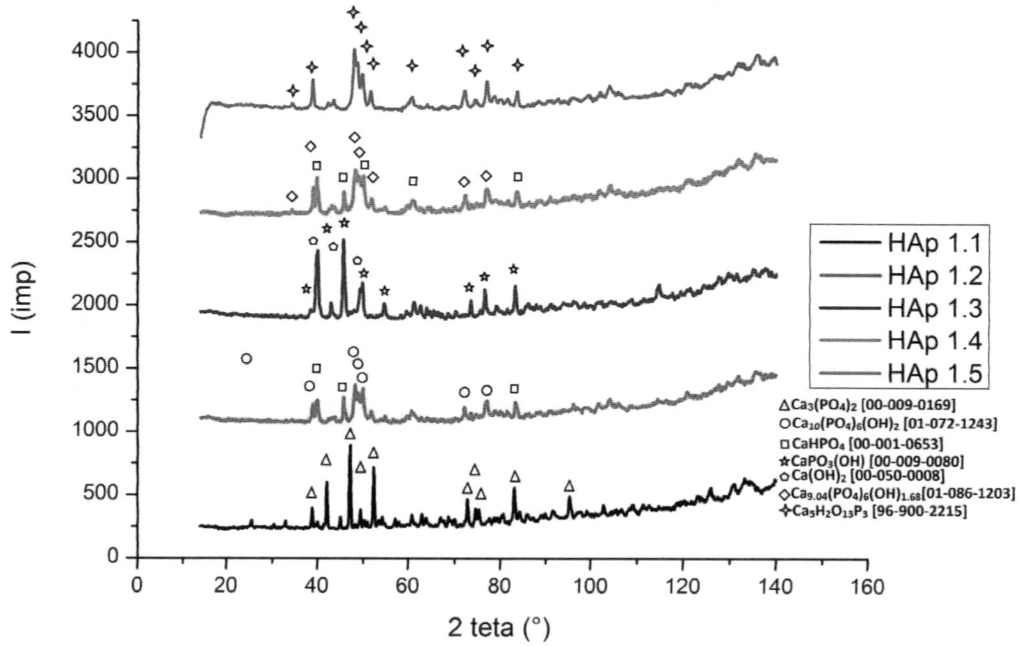

Figure 1. XRD Patterns of the hydroxyapatite samples.

The phase composition of the HAp samples was presented in table 2.

Table 2. HAp samples phase composition.

Sample	Phase composition
HAp 1.1	$Ca_3(PO_4)_2$ [00-009-0169]
HAp 1.2	$CaHPO_4$ [00-001-0653], $Ca_{10}(PO_4)_6(OH)_2$ [01-72-1243]
HAp 1.3	$Ca(OH)_2$ [00-050-0008], $CaPO_3(OH)$ [00-009-0080]
HAp 1.4	$CaHPO_4$ [00-003-0423], $Ca_{9.04}(PO_4)_6(OH)_{1.68}$ [01-086-1203]
HAp 1.5	$Ca_5H_2O_{13}P_3$ [96-900-2215]

The hydroxyapatite crystallinity degree can be analyzed only for the HAp 1.5 sample due to HAp 1.1 and HAp 1.3 haven't hydroxyapatite phase. The HAp 1.2 and HAp 1.4 consists more than one phase. The presence of the second phase makes crystallinity degree analysis impossible due to intensity peaks of the second phase are overlaying on diffraction patterns. The hydroxyapatite samples crystallinity degree was calculated for the sample HAp 1.5 using figure 2. The figure 2 presents the HAp 1.5 remaster XRD pattern without background.

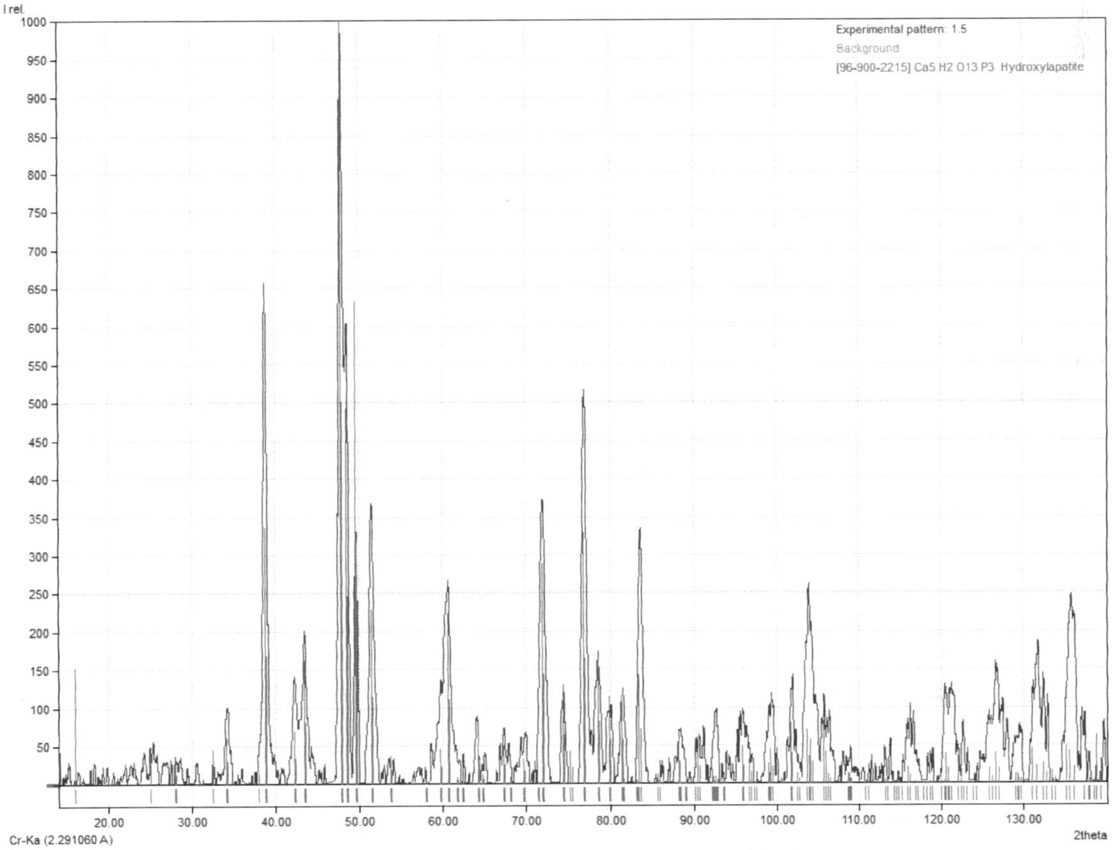

Figure 2. Hydroxyapatite HAp 1.5 XRD pattern without background.

The crystallinity degree calculated via equation (1) was equal to 0.96. This fact provides the ability to use high crystallinity hydroxyapatite as a material for the biomedical application.

The results of FTIR and Raman spectroscopy are presented in figure 3.

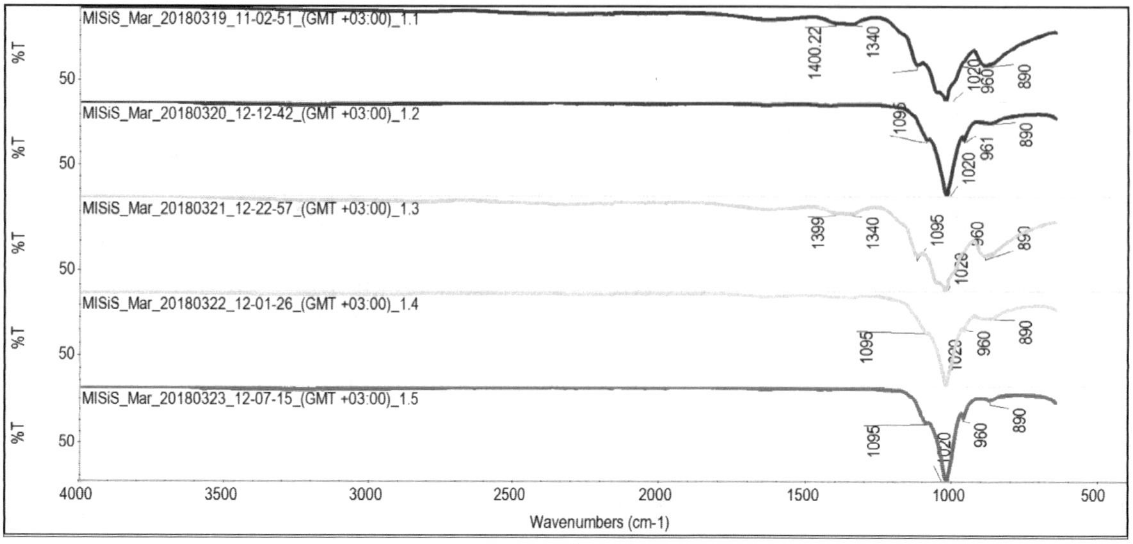

a)

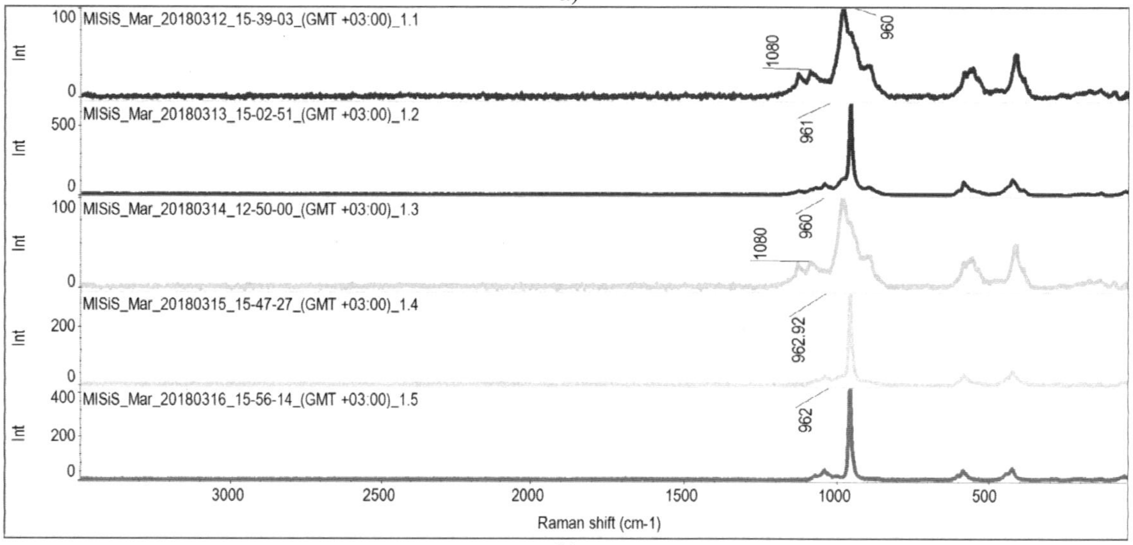

b)

Figure 3. FTIR a) and Raman b) spectroscopy results.

The FTIR specters of the HAp obtained samples demonstrate peaks on 890, 960 1020 and 1095 cm^{-1} reciprocal wavelength. Peak at 890 cm^{-1} refer to vibrations of OH^{-}. Peak at reciprocal wavelength 960 cm^{-1} leads to valence oscillations of phosphate groups (PO_4^{3-}). Peaks at 1020 and 1095 cm^{-1} lead to phosphate group deformation oscillations.

Raman specters show the oscillation mode at 960 cm^{-1} determined as phosphate group.

The HAp 1.1 FTIR specter has additional three peaks which could be described as out-of-plane oscillations of carbonate groups (CO_3^{2-}) and peaks at 1340 and 1400 cm^{-1} described as carbonate group oscillations. The same results are on the FTIR specter of the HAp 1.3 sample.

The FTIR and Raman data support the results of X-Ray diffraction analysis.

The presented X-RAY diffraction FTIR and Raman spectroscopy data show the good perspectives of the $Ca(NO_3)_2$-$(NH_4)_2HPO_4$-NH_4OH precursor system using to obtain the one phase high crystallinity hydroxyapatite powder. Also, the combined chemical precipitation and hydrothermal synthesis method makes possible the hydroxyapatite synthesis. The chemical precipitation at pH level

11 with following hydrothermal synthesis method in stainless steel at 250 °C and 150 bar provide the one phase hydroxyapatite powder with high (0.96) crystallinity degree.

4. Conclusion

The high crystallinity one phase hydroxyapatite $Ca_5H_2O_{13}P_3$ was synthesized used hydrothermal method. The hydroxyapatite samples synthesized using three systems $Ca(OH)_2$-H_3PO_4, $Ca(NO_3)_2$-$(NH_4)_2HPO_4$-NH_4OH, $Ca(OH)_2$-$NH_4H_2PO_4$, shows the perspective of using $Ca(NO_3)_2$-$(NH_4)_2HPO_4$-NH_4OH to obtain one phase high crystallinity hydroxyapatite. The crystallinity was determined at for sample HAp 1.5 was equal to 0.96.

Acknowledgements

The authors acknowledge the financial support of the Ministry of Education and Science of the Russian Federation (Project no. RFMEFI57517X0168).

References

[1] Pu'ad N M, Koshy P, Abdullah H Z, Idris M I and Lee T C 2019 Syntheses of hydroxyapatite from natural sources *Heliyon 5* In press
[2] Okada M and Matsumoto T 2015 Synthesis and modification of apatite nanoparticles for use in dental and medical applications *Jpn. Dent. Sci. Rev.* **51** 85
[3] Okada M 2012 Hydroxylapatite nanoparticles: fabrication methods and medical applications*Sci. Technol. Adv. Mater.* **13** 6
[4] Szcze A, Ho L, Chibowski E 2017 Synthesis of hydroxyapatite for biomedical applications *Adv. Colloid Interface Sci.* 1
[5] G M Raghavendra, K Varaprasad and T Jayaramudu 2015 Chapter 2 - biomaterials: design, development and biomedical applications, in: S. Thomas, Y. Grohens, N. Ninan (Eds.), *Nanotech. Appl. for Tisue Engineering*, William Andrew Publishing, Oxford, 21
[6] N Lertcumfu, P Jaita, S Manotham, P Jarupoom, S Eitssayeam, K Pengpat and G Rujijanagul 2016 Properties of calcium phosphates ceramic composites derived from natural materials *Ceram. Int.* 10638
[7] M Sadat-Shojai, M T Khorasani, E Dinpanah-Khoshdargi and A Jamshidi 2013 Synthesis methods for nanosized hydroxyapatite with diverse structures *Acta Biomater*. **9** 7591
[8] A A Hendi 2017 Hydroxyapatite based nanocomposite ceramics *J. Alloy. Comp* **712** 147
[9] P O'Hare, B J Meenan, G A Burke, G Byrne, D Dowling and J A Hunt 2010 Biological responses to hydroxyapatite surfaces deposited via a co-incident microblasting technique *Biomaterials* **31** 515
[10] Teshima K, Lee S H, Sakurai M, Kameno Y, Yubuta K, Suzuki T, et al. 2009 Wellformed one-dimensional hydroxyapatite crystals grown by an environmentally friendly flux method *Cryst Growth Des* **9** 2937
[11] Mochales C, Wilson RM, Dowker SEP, Ginebra MP. 2011 Dry mechanosynthesis of nanocrystalline calcium deficient hydroxyapatite: structural characterisation. *J Alloys Compd*, 509, 7389
[12] Sadat-Shojai M, Khorasani MT, Jamshidi A, Irani S 2013 Nano-hydroxyapatite reinforced polyhydroxybutyrate composites: a comprehensive study on the structural and in vitro biological properties *Mater Sci Eng C* **33** 2776
[13] Sturgeon J L, Brown P W 2009 Effects of carbonate on hydroxyapatite formed from $CaHPO_4$ and $Ca_4(PO_4)_2O$ *J Mater Sci: Mater Med* **20** 1787
[14] Montazeri N, Jahandideh R and Biazar E 2011 Synthesis of fluorapatite–hydroxyapatite nanoparticles and toxicity investigations *Int J Nanomedicine* **6** 197
[15] Sadat-Shojai M, Khorasani M T and Jamshidi A 2012 Hydrothermal processing of hydroxyapatite nanoparticles – a Taguchi experimental design approach *J Cryst Growth* 361 3

[16] Ponomareva N, Poprygina T, Karpov S, Lesovoi M and Agapov B 2010 Microemulsion method for producing hydroxyapatite *Russ J Gen Chem.* 80 905
[17] Rouhani P, Taghavinia N and Rouhani S 2010 Rapid growth of hydroxyapatite nanoparticles using ultrasonic irradiation. *Ultrason Sonochem* **17** 853
[18] Giardina M A and Fanovich M A 2010 Synthesis of nanocrystalline hydroxyapatite from Ca(OH)2 and H3PO4 assisted by ultrasonic irradiation *Ceram Int.* **36** 1961
[19] Zhang J, Zhan X, Wen X, Song B, Ma L and Peng W 2009 Effects of ultrasonic and dispersants on shape and composition of hydroxyapatite by reflux method *Inorg Mater* **45** 1362
[20] Aghayan M and Rodriguez M 2012 Influence of fuels and combustion aids on solution combustion synthesis of Bi-phasic Calcium Phosphates (BCP) *Mater Sci Eng C* **32** 2464
[21] Itatani K, Tsugawa T, Umeda T, Musha Y, Davies I J and Koda S 2010 Preparation of submicrometer-sized porous spherical hydroxyapatite agglomerates by ultrasonic spray pyrolysis technique *J Ceram Soc Jpn* **118** 462
[22] Huang Y C, Hsiao P C and Chai H J 2011 Hydroxyapatite extracted from fish scale: effects on MG63 osteoblast-like cells *Ceram Int* **37** 1825
[23] Zyman Z, Goncharenko A, Rokhmistrov D and Epple M 2011 Nanocrystalline calciumdeficient hydroxyapatite prepared by a microwave-assisted solvent-free reaction. *Mat-wiss u Werkstofftech* **42** 154
[24] Yudin A et al. 2019 Microwave treatment and pH influence on hydroxyapatite morphology and structure *IOP Publishing Journal of Physics: Conference Series* 1145
[25] Chuprunov K et al. 2018 The ultrasound effect on the morphological properties of hydroxyapatite *MATEC Web of Conferences* EDP Sciences 243
[26] Berzina-Cimdina L and Borodajenko N 2012 Research of calcium phosphates using Fourier transform infrared spectroscopy *Infrared Spectroscopy-Materials Science, Engineering and Technology* **12(7)** 251-263

IOP Conf. Series: Materials Science and Engineering **731** (2020) 012024 doi:10.1088/1757-899X/731/1/012024

The Hydrothermal Synthesis Duration Influence on Calcium Phosphate and Hydroxyapatite Phase Composition

K Chuprunov[1], A Yudin[1], D Leybo[1], D Lysov[1], E Kolesnikov[1], D Kuznetsov[1], A Godymchuk[2,3] and A R Mandal[1]

[1]Department of Functional Nanosystems and High-Temperature Materials, National University of Science & Technology "MISIS", 4 Leninsky avenue, Moscow 119049, Russia
[2]Materials Science Department, Tomsk Polytechnic University, 30 Lenina street, Tomsk 634050, Russia
[3]Chemistry Environmental Laboratory, Tobolsk Complex Scientific Station, Ural Branch of the Russian Academy of Science, 15 Akademician Osipov Street, Tobolsk 626152, Russia

E-mail: kch@misis.ru

Abstract. This paper reports the hydrothermal synthesis process duration influence on phase composition, crystallinity degree, morphology and dispersity of the hydroxyapatite powder. The calcium phosphate and hydroxyapatite were synthesized in precursor system $Ca(NO_3)_2$-$(NH_4)_2HPO_4$-NH_4OH. The obtained powders were characterized with X – Ray diffraction, Fourier Transform spectroscopy (FTIR) and Raman spectroscopy. The results demonstrate: the increasing of the synthesis time duration has no influence on the phase composition. However, the synthesis duration time growth from 12 to 48 hours make possible increase the crystallinity degree from 0.68 to 0.98.

1. Introduction

Nowadays biocompatible materials are attracting growing interest due to own unusual properties and applicability in a field of medicine for tissue engineering, fillers for healing defect in bones and drug delivery systems [1]. Calcium phosphates are the wide class of the ceramic materials. The hydroxyapatite ($Ca_{10}(PO_4)_3(OH)_2$) is one the most thermodynamically stable phase of calcium phosphate phases. Human bone material composition could be divided in three material groups: bone tissue as a calcium phosphate/hydroxyapatite 65 – 70 %, water 5 – 8 % and other percentage is the organic components mainly collagen [2]. Collagen provides elastic resistance properties and acts as a matrix for the mineral crystal formation and crystal growth. The native calcium phosphate usually calcium deficient and carbon surplus HAp with the Ca/P element proportion less than 1.67 [3]. For a long time synthetic hydroxyapatite (HAp) has been interest due to high biocompatibility properties [4], affinity to biopolymers [5] and osteogenic possibility [6]. Also, hydroxyapatites are perspective material as a drug delivery system [7].

The morphology, size (dispersity), phase composition, crystallinity degree, surface area and porosity have an important influence on hydroxyapatite mechanical and biomedical properties. The changes in synthesis parameters make possible control of the mechanical and biomedical properties of the hydroxyapatite samples [8].

The solid state [9], mechanochemical [10], chemical precipitation [11], hydrolysis [12], sol-gel [13], hydrothermal [14], emulsion [15], sonochemical [16], combustion [17], pyrolysis methods [18], synthesis from biogenic sources [19], mixed methods [20] are used to synthesize hydroxyapatite with variable properties.

The mission of presented paper was to analyze the phase composition, crystallinity degree, dispersity and morphological properties of the hydroxyapatite samples synthesized with the hydrothermal technique described at papers [21,22] with variable duration of the hydrothermal treatment process.

The hydrothermal treatment duration influence on the phase composition and crystallinity degree of the calcium phosphate and hydroxyapatite powder was first time discussing in the presented paper. The calcium phosphate based powders were synthesized first time via combined chemical precipitation method with followed hydrothermal synthesis method.

2. Materials and Methods

2.1. Hydrothermal synthesis of hydroxyapatite samples

The calcium phosphate and hydroxyapatite samples were synthesized in the precursor system $Ca(NO_3)_2$-$(NH_4)_2HPO_4$-NH_4OH. The calcium nitrate tetrahydrate ($Ca(NO_3)_2 \bullet 4H_2O$) analytical grade reagent (AR Grade) made by Reachem Co. (Russia), diammonium hydrogen phosphate $(NH_4)_2HPO_4$ (AR Grade) made by Reachem Co. (Russia) and ammonium hydroxide (NH_4OH) (AR Grade) made by Reachem Co. (Russia) were used as an initial precursors for the hydroxyapatite synthesis.

The 10 weight % diammonium hydrogen phosphate water solution was added to the 10 weight % calcium nitrate water solution. The ratio Ca/P was calculated as 1.67. The addition of the 10 weight % water solution ammonium hydroxide increased the pH level, controlled with the pH-meter Mettler Toledo MP230, up to 11. Chemicals precursors were constantly mixed with the rotation mechanical mixer Heidolph RZR 2051 at 500 rpm for 2 hours. The chemical reaction provided with the chemical reaction (1).

$$10Ca(NO_3)_2 + 6(NH_4)_2HPO_4 + 8NH_4OH \rightarrow Ca_{10}(PO_4)_6(OH)_2 + 20NH_4NO_3 \quad (1)$$

The obtained suspension was treated with hydrothermal synthesis method. The hydrothermal synthesis was carried out in stainless steel autoclave fettled with polytetrafluorethylene (PTFE) at 250 – 270 °C and 150 – 200 atm. for 12, 24, 36 and 48 hours consequently. The hydroxyapatite samples obtained with the hydrothermal synthesis presented in table 1.

Table 1. The hydroxyapatite samples obtained with hydrothermal method.

Sample	Synthesis condition		
	Precursor system	**pH**	**Treatment duration, h**
HAp 2.1	$Ca(NO_3)_2 - (NH_4)_2HPO_4 - NH_4OH$	11	12
HAp 2.2	$Ca(NO_3)_2 - (NH_4)_2HPO_4 - NH_4OH$	11	24
HAp 2.3	$Ca(NO_3)_2 - (NH_4)_2HPO_4 - NH_4OH$	11	36
HAp 2.4	$Ca(NO_3)_2 - (NH_4)_2HPO_4 - NH_4OH$	11	48

The next step was washing the hydroxyapatite samples in distilled water until pH level goes down to 7. After that, samples were dried out in vacuum dryer Binder ED 23 at 90 °C for 12 hours.

2.2. Analytical methods to study hydroxyapatite samples

The X-Ray diffraction method Scientific Instruments Difrey 401 (Russia) using Cr Kα emission with wave length 2.2909 Å was used to make a diffraction pattern. The diffraction patterns were detected in angles 14 – 140 ° 2θ with interval 58 ° and time of equilibrium 300 second. The PDF-2004 and COD (Crystallography Open Database) were used to determine the diffraction patterns. The equation (2) was used to analyze crystallinity degree (X_c) of the obtained HAp samples. $V_{112/300}$ described the minimum intensity between two reflections from crystal plane (112) and (300) and I_{300} was the intensity reflection for plane (300).

$$X_c = 1 - \frac{V_{112/300}}{I_{300}} \tag{2}$$

The Fourier Transform Infrared spectroscopy (FTIR spectroscopy) Thermo Scientific Nicolet 380 (USA) and Raman Spectroscopy Thermo Scientific SXR microscope (USA) with the OMNIC operation system and database were used to analyze the phase composition of the hydroxyapatite samples. The IF and Raman specters were determined with book [23].

3. Results and discussion

The X-Ray diffraction patterns presented in the figure 1 show the phase composition of hydroxyapatite synthesized with hydrothermal method. All samples have one phase composition: HAp 2.1 - $Ca_{9.04}(PO4)_3(OH)$ [01-068-1203], HAp 2.2 - $Ca_5(PO_4)_3(OH)$ [01-089-4405], HAp 2.3 - $Ca_5(PO_4)_3(OH)$ [96-900-2216] and HAp 2.4 - $Ca_5(PO_4)_3(OH)$ [96-900-2216].

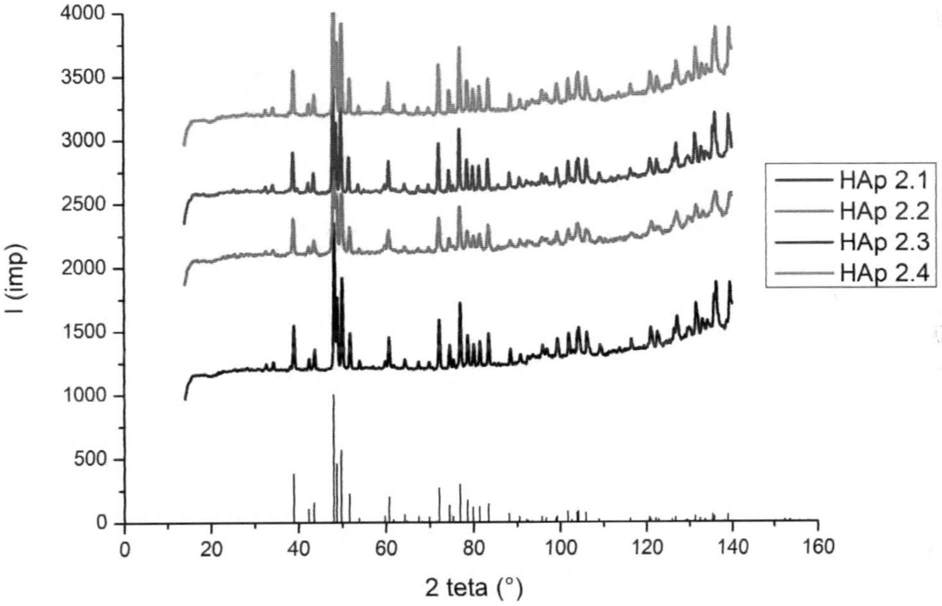

Figure 1. Hydroxyapatite XRD patterns.

The crystallinity degree was calculated with equation 2. The remastered (without background) X-Ray patterns are presented in the figure 2.

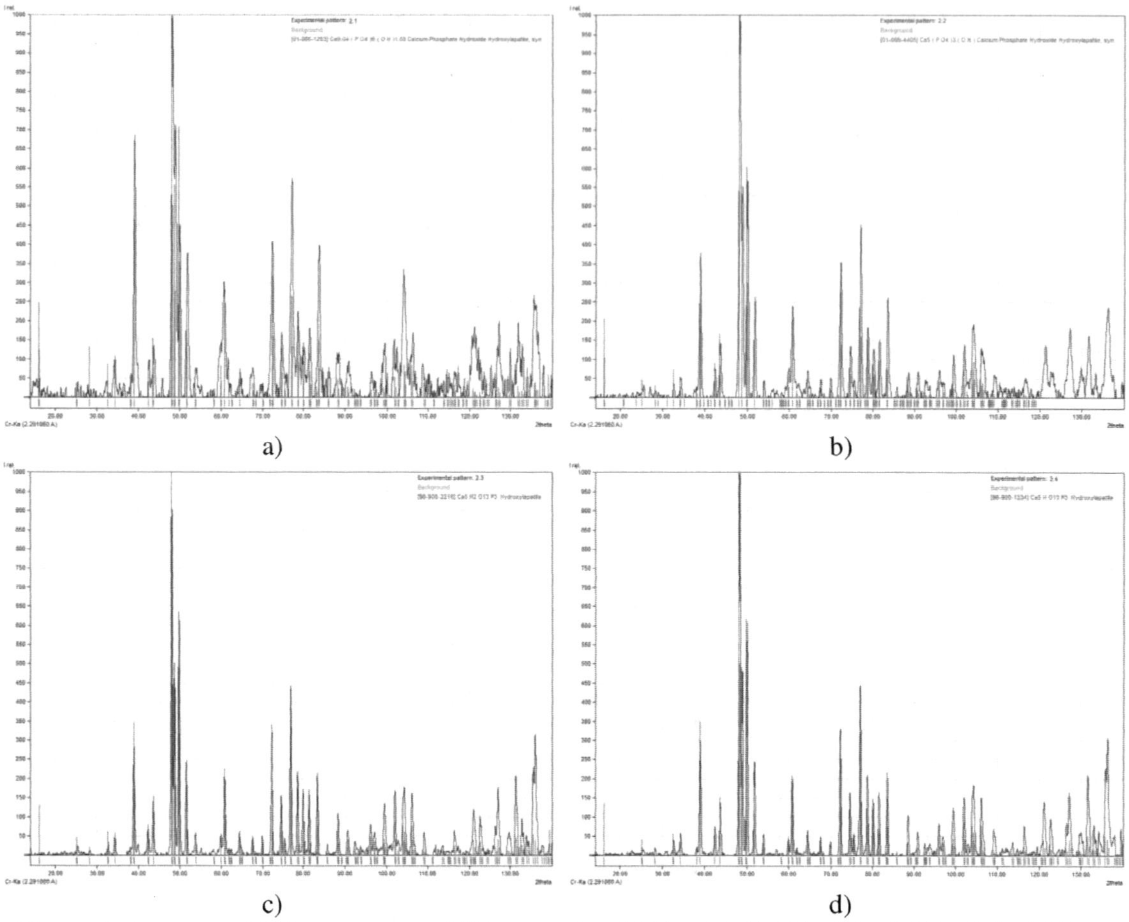

Figure 2. Remastered X-Ray patterns of HAp samples a) HAp 2.1, b) HAp 2.2, c) HAp 2.3, d) HAp 2.4.

The phase composition and crystallinity degree of HAp samples are presented in table 2. The crystallinity degree increases from 0.68 up to 0.98 with the duration hydrothermal synthesis growth from 12 up to 48 hours. The crystallinity degree makes a sharp spike (from 0.68 to 0.98) when time duration increases from 12 to 24 hours. The hydrothermal treatment duration rise from 24 to48 hours hasn't effect on the crystallinity degree of the hydroxyapatite samples. The difference in crystallinity could be described as a measurement error.

Table 2. The phase composition and crystallinity degree of HAp samples.

Sample	Treatment duration, h	Phase composition	Crystallinity degree
HAp 2.1	12	$Ca_{9.04}(PO_4)_3(OH)$ [01-068-1203]	0.68
HAp 2.2	24	$Ca_5(PO_4)_3(OH)$ [01-089-4405]	0.95
HAp 2.3	26	$Ca_5(PO_4)_3(OH)$ [96-900-2216]	0.96
HAp 2.4	48	$Ca_5(PO_4)_3(OH)$ [96-900-2216]	0.98

Figure 3 shows FTIR and Raman spectroscopy results.

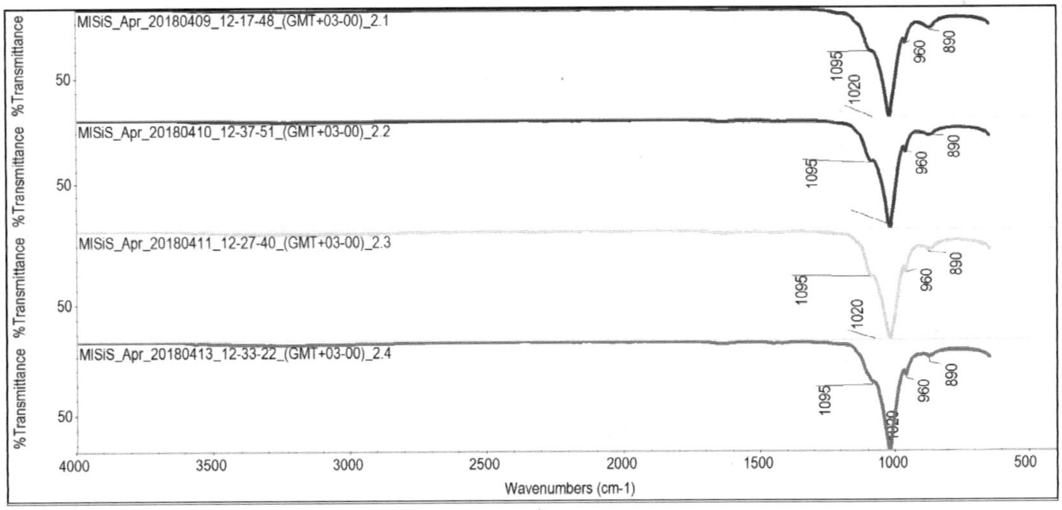

a)

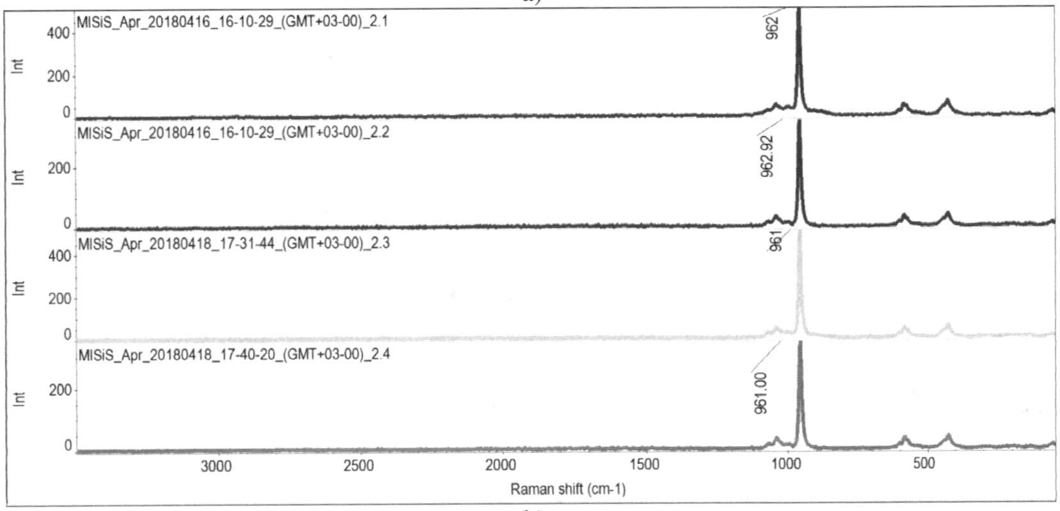

b)

Figure 3. FTIR a) and Raman b) spectroscopy data.

For all samples FTIR specters have 4 lines at 890, 960, 1020 and 1095 cm^{-1}. Peak of intensity at 890 cm^{-1} corresponds for vibrations of the OH^- hydroxide group. Line at 960 cm^{-1} responses for valence vibrations phosphate group (PO_4^{3-}), and lines at 1020 and 1095 cm^{-1} refer to phosphate group deformation vibrations.

Line at 960 cm^{-1} at Raman specter corresponds to fundamental vibrations of phosphate groups PO_4^{3-} and it was the most intensive line on all the specters.

The FTIR and Raman data support the results of X-Ray diffraction analysis.

The presented X-Ray diffraction, FTIR and Raman spectroscopy datas show the good perspectives of the combined chemical precipitation with the followed hydrothermal synthesis method to produce hydroxyapatite powders. The obtained data demonstrate the chemical precipitation in a precursor system $Ca(NO_3)_2$-$(NH_4)_2HPO_4$-NH_4OH at pH level 11 with followed hydrothermal treatment at 250 °C and 150 bar provide synthesis of the one phase hydroxyapatite powder.

The crystallinity degree analysis demonstrates the increasing of the crystallinity degree from the 0.68 up to 0.95 with the growth of the hydrothermal treatment duration from 12 to 24 hours. The follow increasing of the hydrothermal treatment duration from 24 to 48 hours has insignificant effect on the crystallinity degree (X_c (HAp 2.2) = 0.95, X_c (HAp 2.3) = 0.96 and X_c (HAp 2.4) = 0.98). The difference between crystallinity degree can be described as the measurement error.

4. Conclusion

The hydroxyapatite synthesized with hydrothermal method at different time duration (12, 24, 36 and 48 hours) at 250 – 270 °C and 150 – 200 atm using precursor system $Ca(NO_3)_2$-$(NH_4)_2HPO_4$-NH_4OH shows one phase composition. Primary, all samples consist of the $Ca_5(PO_4)_3(OH)$ phase for all obtained HAp samples. The FTIR and Raman results support the X-Ray analysis data. The crystallinity degree sharply increases from 0.68 to 0.95 with the growth of the time duration of hydrothermal synthesis from 12 to 24 hours. The time duration of the hydrothermal treatment process increases from 24 to 48 hours slightly rises the crystallinity degree of the HAp samples up to 0.98.

Acknowledgements

The authors acknowledge the financial support of the Ministry of Education and Science of the Russian Federation (Project no. RFMEFI57517X0168).

References

[1] Marten A, Fratzl P, Paris O and Zaslansky P 2010 *Biomaterials* 31 5479

[2] Sadat-Shojai M., Khorasani M T, Dinpanah-Khoshdargi E and Jamshidi A 2013 *Acta biomaterialia 9* 7591

[3] Rabiei A, Blalock T, Thomas B, Cuomo J, Yang Y and Ong J 2007 Microstructure, mechanical properties, and biological response to functionally graded HA coatings *Mater Sci Eng C* 27 529

[4] Chen L, Mccrate JM, Lee JCM and Li H 2011 *Nanotechnology* 22 105708

[5] Pelin I M, Maier S S, Chitanu G C and Bulacovschi V 2009 Preparation and characterization of a hydroxyapatite–collagen composite as component for injectable bone substitute. *Mater Sci Eng C* 29 2188

[6] O'Hare P, Meenan BJ, Burke GA, Byrne G, Dowling D and Hunt JA 2010 Biological responses to hydroxyapatite surfaces deposited via a co-incident microblasting technique *Biomaterials* 31 515

[7] Chen S, Shi Y, Luo Y and Ma J 2019 Layer-by-layer coated porous 3D printed hydroxyapatite composite scaffolds for controlled drug delivery *Colloids and Surfaces B: Biointerfaces* 179 121

[8] Bose S and Tarafder S 2012 Calcium phosphate ceramic systems in growth factor and drug delivery for bone tissue engineering: a review *Acta biomaterialia* 8 4 1401

[9] Pramanik S, Agarwal AK, Rai KN and Garg A 2007 Development of high strength hydroxyapatite by solid-state-sintering process *Ceramics International* 33 (3) 419

[10] Mochales C, Wilson RM, Dowker SE and Ginebra MP 2011 Dry mechanosynthesis of nanocrystalline calcium deficient hydroxyapatite: Structural characterization *Journal of alloys and compounds* 509 (27) 7389

[11] Zhang Y and Lu J 2008 The transformation of single-crystal calcium phosphate ribbon-like fibres to hydroxyapatite spheres assembled from nanorods *Nanotechnology* 19 (15) 155608

[12] Seo DS and Lee JK 2008 Synthesis of hydroxyapatite whiskers through dissolution–reprecipitation process using EDTA *Journal of Crystal Growth* 310 (7-9) 2162

[13] Padmanabhan SK, Balakrishnan A, Chu MC, Lee YJ, Kim TN and Cho SJ 2009 Sol–gel synthesis and characterization of hydroxyapatite nanorods *Particuology* 7 (6) 466

[14] Zhang G, Chen J, Yang S, Yu Q, Wang Z and Zhang Q 2011 Preparation of amino-acid-regulated hydroxyapatite particles by hydrothermal method *Materials letters* 65 (3) 572

[15] Pradeesh TS, Sunny MC, Varma HK and Ramesh P 2005 Preparation of microstructured hydroxyapatite microspheres using oil in water emulsions *Bulletin of Materials Science 28*(5) 383

[16] Han Y, Wang X and Li S 2009 A simple route to prepare stable hydroxyapatite nanoparticles suspension *Journal of Nanoparticle Research* 11 (5) 1235

[17] Ghosh S K, Pal S, Roy SK, Pal S K and Basu D 2010 Modelling of flame temperature of solution combustion synthesis of nanocrystalline calcium hydroxyapatite material and its parametric optimization *Bulletin of Materials Science* 33(4) 339

[18] Cho JS and Kang YC 2008 Nano-sized hydroxyapatite powders prepared by flame spray pyrolysis *Journal of Alloys and Compounds* 464 (1-2) 282

[19] Gergely G, Wéber F, Lukács I, Illés L, Tóth A L, Horváth Z E and Balázsi C 2010 Nano-hydroxyapatite preparation from biogenic raw materials *Central European Journal of Chemistry* 8 (2) 375

[20] Abdel-Aal E A, El-Midany A A and El-Shall H 2008 Mechanochemical–hydrothermal preparation of nano-crystallite hydroxyapatite using statistical design *Materials Chemistry and Physics* 112 (1) 202

[21] Yudin A et al. 2019 Microwave treatment and pH influence on hydroxyapatite morphology and structure *IOP Publishing Journal of Physics: Conference Series* 1145

[22] Chuprunov K et al. 2018 The ultrasound effect on the morphological properties of hydroxyapatite. *MATEC Web of Conferences EDP Sciences* 243

[23] Berzina-Cimdina L and Borodajenko N 2012 Research of calcium phosphates using Fourier transform infrared spectroscopy *Infrared Spectroscopy-Materials Science, Engineering and Technology* 12 (7) 251-263

Effect of Parameters of Ultrasonic Shot Peening on the Physicomechanical Properties of the Surface Layer of Flat Samples

A A Shiryaev[1], V N Trofimov[1], V V Karmanov[1] and D V Vindokurov[2]

[1]Perm National Research Polytechnic University, 614990 29 Komsomolsky prospekt, Perm, Russian Federation
[2]UEC-Perm Engines. 614010 93 Komsomolsky prospekt, Perm, Russian Federation

alex_sh_23-1@mail.ru

Abstract. The work is devoted to the study of the influence of the parameters of ultrasonic shot peening on the physicomechanical properties of the surface layer of flat samples of 08ps steel, AMg2 and M2 alloys. The results of determination of residual stresses, microhardness are given. As a result, it was established that the processing time makes the largest contribution in the selected modes.

1. Introduction

The use of surface plastic deformation (SPD) methods at the final stages of manufacture allows to improve the operational properties of parts and thereby increase the service life of machines. SPD methods allow you to process parts of various configurations and at the same time ensure high quality of the surface layer and high productivity of processing processes [1-2].

Physico-mechanical properties of the surface of parts determine their reliability during operation [3-4]. The main characteristics of the physico-mechanical properties of the surface of the parts are:

- depth of the hardened layer (hardening depth);
- level and sign of residual stresses;
- hardness.

One of the reasons for premature failure of parts is tensile residual stresses (RS). Therefore, the urgent task is to create a compressive RS in the surface layer, contributing to an increase in fatigue strength.

As a result of SPD, the surface roughness decreases, the hardness (microhardness) of the metal increases, and compressive residual stresses appear in the surface layer of the part.

The most effective and common SPD processes include [5]:

- bead-blasting treatment (pneumatic blasting);
- hydroblasting treatment;
- ultrasonic shot peening.

The effect of the parameters of ultrasonic bead-blasting processing on the physicomechanical properties of the surface layer of flat samples (Fig. 1) from 08ps steel, aluminum alloy AMg2, copper M2 are investigated.

2. Method of the experiment. Samples and equipment

Sample materials - steel 08ps, alloys AMg2, M2. Sizes of the samples, mm - 70 x 19 x 2 (Fig. 1). The initial roughness of the samples is Ra 1.25. Processing was carried out on the UZU-6TM installation. Ultrasonic shot peening (USP) modes during the experiment:

- waveguide oscillation frequency - 17.50 and 19.35 kHz;
- working bodies - a mixture of metal balls (table 1);
- the processing time of the samples is shown in table 2. USP modes are given in table 2

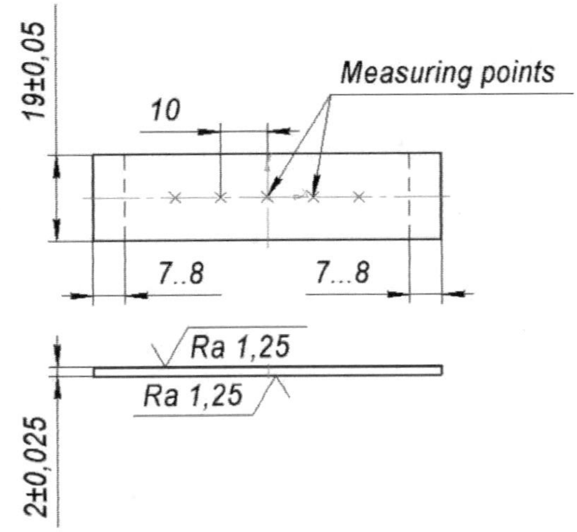

Figure 1. Sketch plate with control points.

Table 1. The composition of the mixture of balls for USP.

Mixture of balls	Diameter of balls (mm)	The mass of balls in the mixture (g)
1	0.68	20
	0.68	5
2	1.00	15
	1.58	25

Table 2. USP modes for 08ps steel and AMg2, M2 alloys.

Sample	Processing time (s)	Frequency (kHz)	Mix
1	20/5	19	1
2	25/10	19	1
3	30/15	19	1/2
4	35/20	19	1/2
5	20/5	17.5	2
6	25/10	17.5	2
7	30/15	17.5	2
8	35/20	17.5	2
9	20/5	17.5	1
10	25/10	17.5	1
11	30/15	17.5	1
12	35/20	17.5	1

Micro roughness was measured with a 5.6 mm MARSURF PS1 instrument. Residual stresses were determined by X-ray diffractometry using an XStress 3000 Robot complex [6].

The microhardness of the surface of the samples according to Vickers was measured with a Shimadzu HMV-G instrument (and $HV_{0.01}$); it was measured at 5 control points before and after the ultrasound examination. The value of $HV_{0.3}$ was determined on the surface of the samples.

When plotting the distribution of microhardness and RS over depth, the samples were etched at point 1 with a step of 10 μm. The diameter of the etching spot is 10 mm. Equipment Used: Movipol-5. The microhardness $HV_{0.01}$ was determined by the average value from the results of 5 measurements in the etching spot at each etching stage. The choice of load is due to the requirement that the diagonal of the print does not exceed the depth of the etched layer.

3. Analysis of the experimental results

The average microhardness of the surface of specimens from 08ps steel to USP is 6 HV, and for specimens from AMg2 and M2 alloys - 3.7 HV and 5 HV, respectively.

3.1. Microhardness

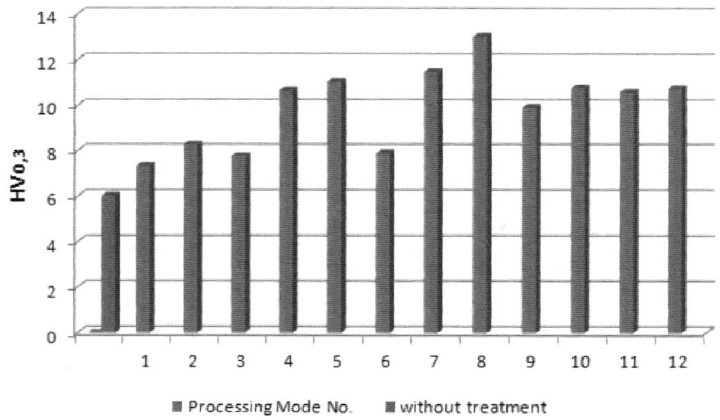

Figure 2. The influence of USP on microhardness for 08ps steel.

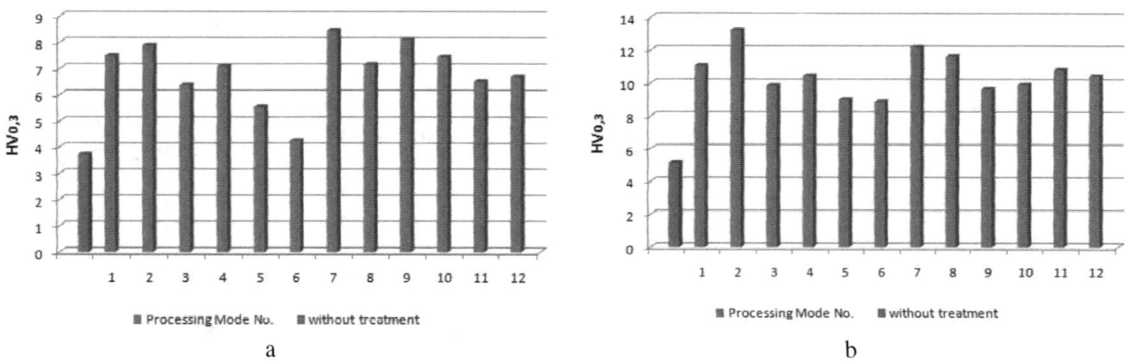

Figure 3. The influence of USP on microhardness: a) AMg2; b) M2.

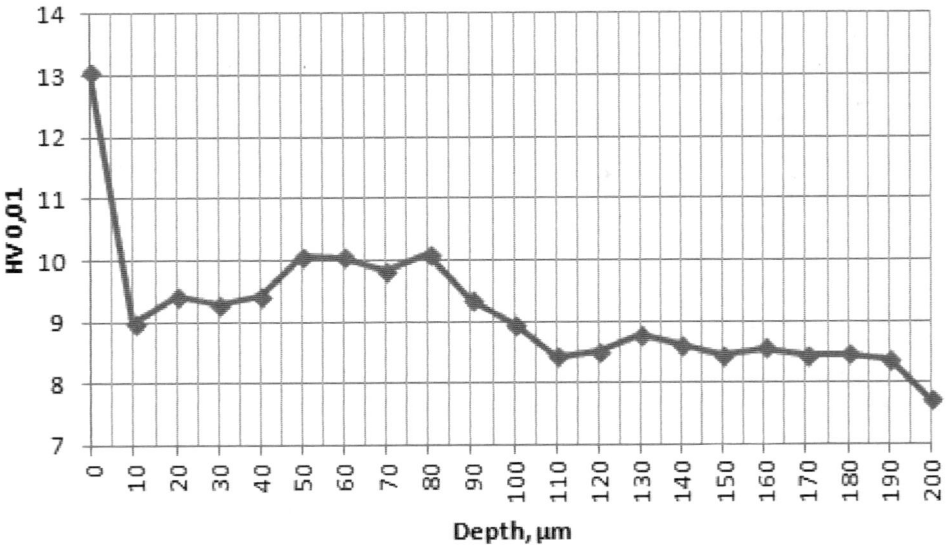

Figure 4. Depth distribution of microhardness.

For the specimen made of 08ps steel with the highest microhardness (sample 8), a diagram of the microhardness distribution over depth was constructed (Fig. 4). A sharp decrease in microhardness is observed at a depth of 10 µm from 13.04 to 8.98. Further, the HV value increases to a depth of 80 µm, after which a gradual decrease to 7.71.

At a depth of 200 µm, the microhardness exceeds the initial microhardness, which indicates that the depth of the hardened layer exceeds the depth of the layer studied by the authors.

3.2. Residual stress

The effect of the USP on the residual stresses measured along the axis of the samples is shown in Figures 5-7.

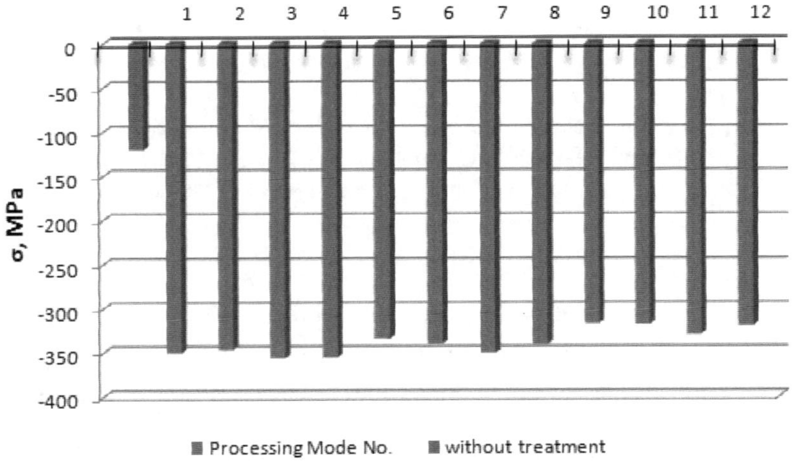

Figure 5. The influence of the USP modes on the level of RS for 08ps steel.

IOP Conf. Series: Materials Science and Engineering **731** (2020) 012025 doi:10.1088/1757-899X/731/1/012025

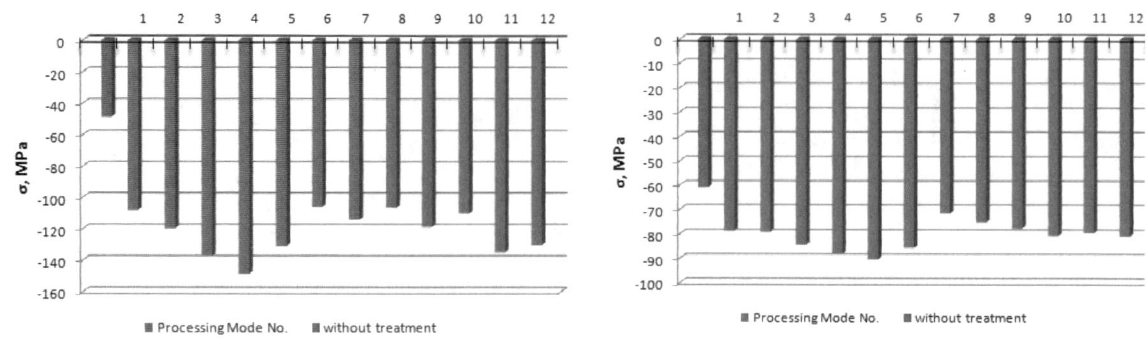

a б

Figure 6. The influence of the USP modes on the RS level: a) AMg2; b) M2.

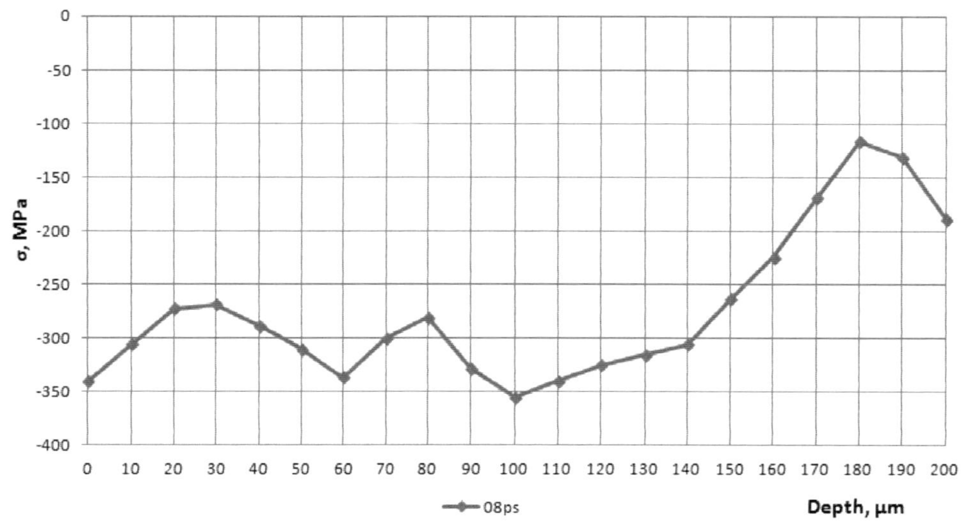

Figure 7. Depth distribution diagram of RS.

For the sample with the highest microhardness (steel 08ps - sample 8), a plot of the distribution of OH over depth is plotted (Fig. 7). The etching modes are the same as in the determination of hardness.

In the surface layer of the 08ps steel specimen, the residual stresses increase and vary in the range of -350 ÷ -120 MPa.

4. Conclusions

Table 3 shows the comparative results of the experiment on the influence of USP on the formation of physical and mechanical properties and surface geometry of the samples.

Table 3. The effect of USP on the surface properties of samples.

Sample Condition	Steel 08ps		AMg2 alloy		M2 alloy	
	HV	RS level, MPa	HV	RS level, MPa	HV	RS level, MPa
Before USP	5,9	-120	3,7	-50	5	-60
After USP	8÷15	-320÷-410	4÷8	-106÷-148	9÷13	-71÷-90

The results show that:

- The composition of the mixture of balls for USP also affects the formation of surface properties of the samples, but to a lesser extent than the processing time.
- The microhardness of the surface of the samples after USP increases: for steel 08ps - by 1.3–2.5 times; for the AMg2 alloy - 1.08 ÷ 2.16 times, for the M2 alloy - 1.8 ÷ 2.6 times.
- A field of compressive RS is formed in the surface layer of the samples after the SPD, the value of which decreases.

General conclusion:

- The use of ultrasonic testing has a beneficial effect on the physical and mechanical properties of the surface of the samples, which leads to an increase in fatigue and contact strength, corrosion resistance.
- The USP method can be recommended for widespread adoption in the processing of parts that are subject to increased reliability requirements for the operation of engineering products

References

[1] Zaydes S A and Kiong N K 2017 The influence of the kinematics of surface plastic deformation processes on the quality of the hardened surface layer *Hardening Technologies and Coatings, Electronic Materials* 13 **5** pp 235-240

[2] Papshev D D 1978 *Finishing and hardening by surface plastic deformation* Moscow: Mashinostroenie p 152

[3] Vishnyakov Ya D and Piskarev V D *Residual stress control in metals and alloys* Moscow Metallurgiya p 254

[4] Suslov A G 2000 *The quality of the surface layer of machine parts* Moscow: Mashinostroenie p 320

[5] Lebedenko V G 2008 Mathematical modeling of the formation of geometric parameters of the surface layer and hardening parameters when processing parts by shot *Vestnik of the Don Technical University, Electronic Materials* 8 **4** pp 202-212

[6] Vasil'ev D M 1977 *Diffraction research methods* Moscow: Metallurgiya p 248

[7] Trofimov V N, Karmanov V V, Shiryaev A A and Zvonov S N 2019 Calibration of X-Ray Diffraction Instruments for Residual-Stress Measurement *Russian Engineering Research, Electronic Materials* 39 **3** pp 276–278

Research Methodology of Amplitude-Frequency Characteristics of Dental (Endodontic) Files

A V Avilov[1] and E V Laktionov[2]

[1]Don State Technical University, 1 Gagarin Square, Rostov-on-Don, Russia
[2]Rostov State Medical University, 119 Suvorov Street, Rostov-on-Don, Russia

av100982@rambler.ru, laktionov.egor@ya.ru

Abstract. Deformation and fracture of rotating nickel-titanium endodontic files is considered to be a serious problem among dentists. Therefore, the study of the causes of microcracks, the study of the stress-strain state of endodontic files, the development of methods for monitoring their performance are an urgent problem for dentistry. The purpose of this article is to describe the method of detection of functional endodontic files, based on the determination of their amplitude-frequency characteristics. A block diagram of the measuring complex for the study of free oscillations of endodontic files is proposed. The results of experimental studies are presented.

1. Introduction

The purpose of endodontic treatment is to clean and shape the root canals of the tooth. The quality of endodontic treatment is important for increasing the life time of the tooth and for successful prosthetic dentistry and therapy. The problem of endodontic treatment is a peculiar anatomy and complex the geometry of the root canal: small size, significant curvature. To facilitate the treatment of curved root canals, nickel-titanium endodontic tools (files) with different geometric characteristics of cross-section, taper and length are used [1]. Despite their increased flexibility, there are unexpected fractures of these files due to cyclic fatigue, increased moments of resistance, the occurrence of microcracks, etc. [2-5]. Especially often file fractures occur after misuse, reuse and sterilization, which may change the mechanical properties of files [6]. After the fracture of the file, part of it remains blocked in the root canal which requires additional efforts to remove it, for example, with the help of ultrasonic vibration and the use of a dental operating microscope [7]. Figure 1 shows the image of a tooth with the remainder of the endodontic file in the dental canal. To prevent fracture of files in the tooth canal, a method of preliminary control of endodontic files by their amplitude-frequency characteristics is proposed.

2. Methodology of determination of amplitude-frequency characteristics of endodontic files

To determine the amplitude-frequency characteristics of a dental instrument of low length and thickness, you can use several methods to obtain the natural oscillation frequencies of endodontic files: use the acoustic radiation, contact methods (for example, using vibroaccelerometers). The proposed technique uses the acoustic radiation method for the control of endodontic files by modes of own oscillations before primary use, after use and sterilization. The essence of the method is to change the amplitude-frequency characteristics of endodontic files during elongation, formation of microfractures and fracture of the instrument.

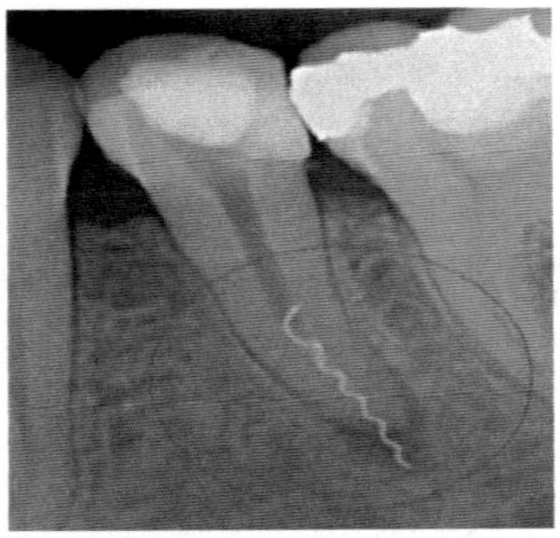

Figure 1. Image of a tooth with the remainder of the endodontic file in the dental canal.

At short-term action of force F on the top part of a cutting edge endodontic file occur rapidly attenuated high frequency oscillations. A sensitive Behringer microphone and sound card SB Live! Platinum 5.1 is used to record the endodontic file's natural oscillation with built-in ADC with high sampling rate up to 96 kHz. A block diagram of natural oscillation frequency measurement of endodontic files is presented in Figure 2. Digital processing and filtering of the acoustic signal was carried out in a free and multi-platform audio editor Audacity with a wide range of functionality for processing audio files.

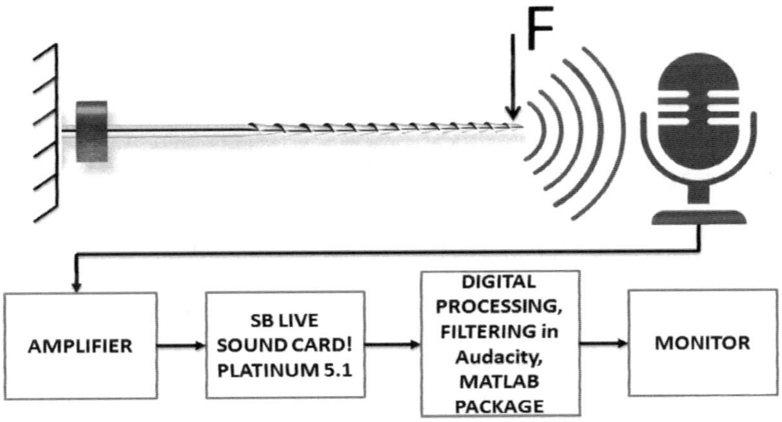

Figure 2. Block diagram of natural oscillation frequency measurement of endodontic files.

Acoustic signals were recorded from new, and endodontic files ProFile 25mm 0.4 A011N 150-300min^{-1} were used. Acoustic signals and natural oscillation frequencies of endodontic files are presented in Figures 3 and 4. A block diagram of the complex for recording amplitude-frequency characteristics of endodontic files is presented in Figure 5.

Endodontic files are installed in the holes of the turntable. The rotation of the table is carried out by the electric drive according to the given program. When the table is rotated, the top of the endodontic file touches the plate rigidly fixed below the upper part of the endodontic file by 3 mm, and begins to make high-frequency oscillations. The acoustic signal is amplified, filtered, and processed. The result of digital signal processing (natural frequency) is displayed on the monitor screen. The digital high-speed camera allows you to photograph and record the vibrations of the endodontic file with the subsequent processing of photographs to determine the frequencies of vibrations. To improve the

IOP Conf. Series: Materials Science and Engineering **731** (2020) 012026 doi:10.1088/1757-899X/731/1/012026

accuracy of research in the complex simultaneously used two methods of registration oscillations of the endodontic file: acoustic and high-speed shooting.

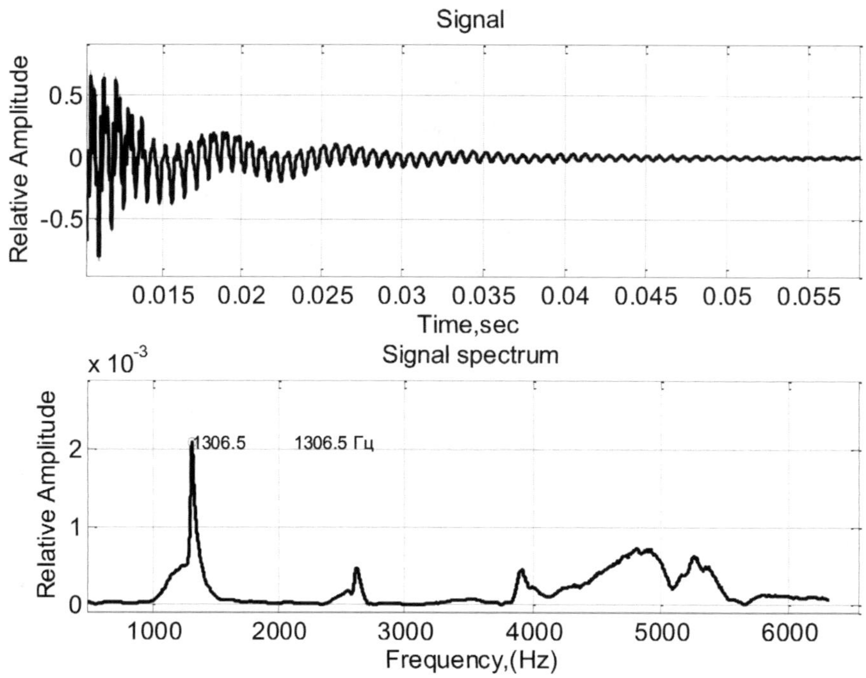

Figure 3. Acoustic signals and natural oscillation frequencies of new endodontic files.

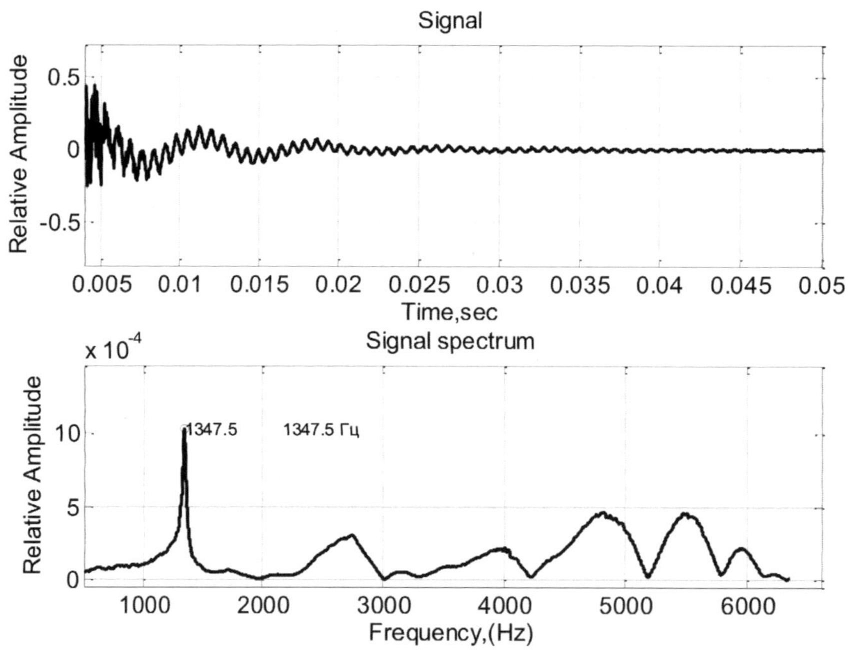

Figure 4. Acoustic signals and natural frequencies of used endodontic files.

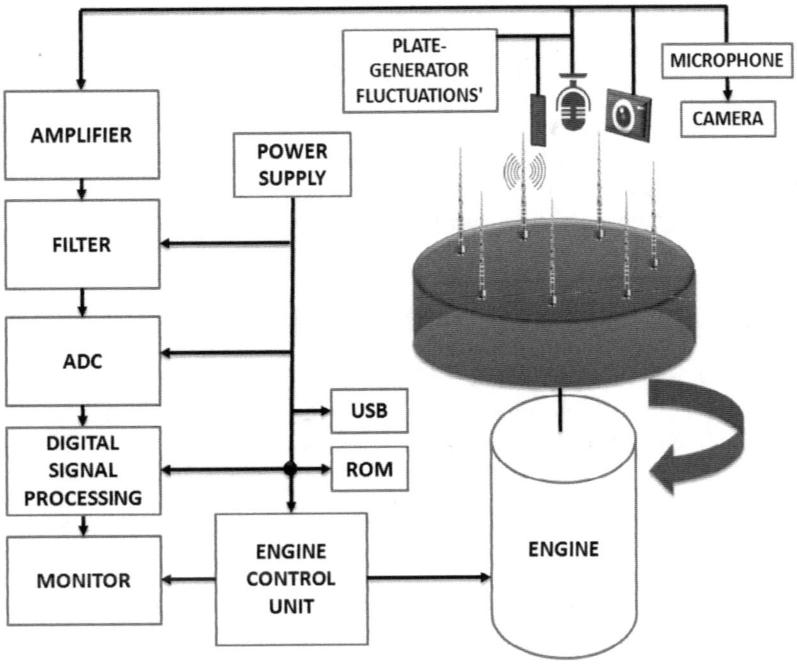

Figure 5. A block diagram of the complex for recording amplitude-frequency characteristics of endodontic files.

3. Results of research

Amplitude-frequency characteristics of 50 new and 25 used endodontic files ProFile 25mm 0.4 A011N 150-300min^{-1} were investigated. When processing the results of the research, the signal received from the microphone contained interference from foreign sources (industrial interference, interference from broadcasting stations, atmospheric interference, etc.). Therefore, the received signals were filtered, processed useful signals, forms and amplitude-frequency characteristics of which are presented in Figures 3 and 4. Figure 3 shows an example of a useful signal and amplitude-frequency response of a new endodontic file. The natural oscillation frequency of the new file is 1306.5 Hz. Figure 4 shows an example of a useful signal and amplitude-frequency response of the endodontic file used with geometry violation. The natural oscillation frequency of the used file is 1347.5 Hz.

When conducting preliminary experiments and developing methods for studying the amplitude-frequency characteristics of endodontic files, it was established:

– for pre - processing of acoustic signals and search for methods of their research, it is convenient to use the extensive capabilities of various audio editors, such as Audacity, as well as extensive signal processing capabilities in the Matlab package;

– the most successful use of automated software module based on the open Python programming language (WAVE), enables immediate automated processing of the data received directly from the microphone;

– the value of the natural oscillation frequency of the endodontic files use is higher than the frequency of new files, which indicates the beginning of the structure changing process of the endodontic file; such files have the probability of fracture in the tooth canal;

– the proposed method for determining the natural oscillation frequencies of endodontic files can be used to control defects in the surface and structure of endodontic files.

References

[1] Nagi S E, Khan F R and Rahman M 2016 Comparison of fracture and deformation in the rotary endodontic instruments: Protaper versus K-3 system *J. Pak Med Assoc.* Mar 66(Suppl 3)(10) pp S30-S32 PubMed PMID 27895347

[2] Wu J, Lei G, Yan M, Yu Y, Yu J and Zhang G 2011 Instrument separation analysis of multi-used ProTaper Universal rotary system during root canal therapy *J. Endod.* Jun 37(6) pp 758-63 PubMed PMID 21787484

[3] Gao M J, Zhou J, Jiang X Z and Liang L 2012 Evaluation of the fracture of nickel-titanium rotary Protaper during root canal preparation *J. Shanghai Kou Qiang Yi Xue.* Jun 21(3) pp 294-7 PubMed PMID 22885490

[4] Farmakis E T, Sotiropoulos G G, Pantazis N and Kozyrakis K 2013 The permanent deformation of the self-adjusting files when used in canals of extracted teeth. *Int Endod J.* Sep 46(9) pp 863-9 PubMed PMID 23442046

[5] Di Fiore P M, Genov K A, Komaroff E, Li Y, Lin L 2006 Nickel-titanium rotary instrument fracture: a clinical practice assessment. *Int Endod J.* Sep 39(9) pp 700-8 PubMed PMID 16916359

[6] Kuhn G and Jordan L. 2001 Microstructural effects on Ni-Ti endodontic instruments failure *J. Phys IV.* Volume 11 Number PR8 pp Pr8-553-8 1051/jp4 2001892

[7] Shahabinejad H, Ghassemi A, Pishbin L and Shahravan A 2013 Success of ultrasonic technique in removing fractured rotary nickel-titanium endodontic instruments from root canals and its effect on the required force for root fracture. *J. Endod.* Jun 39(6) pp 824-8 PubMed PMID 23683286

AUTHOR INDEX

Abdulmenova, E V 82
Abzaev, Yu A 70
Adamov, A A 35
Anoshkin, A N 41, 59, 65
Antonova, L V 16
Ashikhmin, V A 41
Avdeeva, E S 47
Avilov, A V 150
Badin, A V 76, 94, 100
Barbarash, L S 16
Bataev, I A 53
Bolbasov, E N 118, 124
Bouznik, V M 124
Brezina, J 24
Brezinová, J 24
Burkov, M V 1, 6
Burov, V G 53
Buznik, V M 118
Byakov, A V 16
Chuprunov, K 112, 130, 137
Dorofeev, I O 100
Emurlaev, K I 53
Eremin, A V 1, 6, 11, 16
Ermakov, D A 59, 65
Gering, M O 76
Glushkova, T V 16
Godymchuk, A 130, 137
Gomorova, J F 105
Gusev, A 112
Ilinykh, I 130
Ivanov, Yu F 70, 82
Kalashnikov, M P 82
Kamenskikh, A A 35, 47
Karmanov, V V 144
Karmanova, A E 53
Klopotov, A A 70
Kolesnik, I M 118
Kolesnikov, E 130, 137
Kondratyuk, A A 70
Kozulin, A A 11
Krivkina, E O 16
Kuleshov, G E 76
Kuznetsov, D 112, 130, 137
Laktionov, E V 150
Leonov, A A 82
Leybo, D 130, 137
Lysov, D 130, 137
Lyubutin, P S 11
Mandal, A R 137
Maruschak, P 24
Nikulina, A A 53
Nosov, Yu O 35
Panin, S V 11, 16, 105
Panin, S 24
Petrikova, E A 70
Pidotova, D A 94
Pisarev, P V 41, 59, 65
Ruktuev, A. A. 53
Sbrodov, A V 89
Shiryaev, A A 144
Shlyakhova, G 30
Shugurov, V V 82
Silnikov, V N 16
Smygalina, P P 94
Surikova, N S 105
Teresov, A D 70, 82
Teterina, D D 100
Titkov, V V 11
Tolkachev, A S 70
Trofimov, E A 94
Trofimov, V N 144
Tverdokhlebov, S I 118
Tverdokhlebova, T S 124
Ustinov, A M 70
Vindokurov, D V 144
Viňáš, J 24
Vlasov, I V 105
Yakovlev, A V 105
Yntymakova, A 30
Yudin, A 112, 130, 137
Zakharova, O 112